ENERGY AND INFRASTRUCTURE

Volume 6

# Biomass Assessment

# Full list of titles in the set
## ENERGY AND INFRASTRUCTURE

# Biomass Assessment

*Andrew Millington and John Townsend*

from Routledge

First published by Earthscan in the UK and USA in 1989

For a full list of publications please contact:
Earthscan
2 Park Square, Milton Park, Abingdon, Oxfordshire OX14 4RN
711 Third Avenue, New York, NY 10017

First issued in paperback 2016

*Earthscan is an imprint of the Taylor & Francis Group, an informa business*

ISBN 13: 978-1-138-96472-3 (pbk)
ISBN 13: 978-1-84407-978-0 (hbk)

ISBN 978-1-84407-972-8 (Energy and Infrastructure set)
ISBN 978-1-84407-930-8 (Earthscan Library Collection)

Earthscan publishes in association with the International Institute for Environment and Development

A catalogue record for this book is available from the British Library

Library of Congress Cataloging-in-Publication Data has been applied for

**Publisher's note**
The publisher has made every effort to ensure the quality of this reprint, but points out that some imperfections in the original copies may be apparent.

At Earthscan we strive to minimize our environmental impacts and carbon footprint through reducing waste, recycling and offsetting our $CO_2$ emissions, including those created through publication of this book. For more details of our environmental policy, see www.earthscan.co.uk.

# BIOMASS ASSESSMENT

**Woody Biomass in the SADCC Region**

# BIOMASS ASSESSMENT

## Woody Biomass in the SADCC Region

**A study commissioned by the SADCC Energy Secretariat and carried out by the ETC Foundation (Consultants for Development)**

ANDREW MILLINGTON

JOHN TOWNSHEND
PAM KENNEDY
RICHARD SAULL
STEVE PRINCE
ROBERT MADAMS

EARTHSCAN PUBLICATIONS LTD
IN ASSOCIATION WITH THE ETC FOUNDATION
LONDON

First published in 1989 by
Earthscan Publications Ltd
3 Endsleigh Street
London WC1H ODD

*British Library Cataloguing in Publication Data*

Millington, Andrew
  Biomass assessment in the SADCC region.
  1. Africa south of the Sahara. Energy resources.
  Biomass. Distribution
  I. Title II. Townshend, John
  333.79

  ISBN 1-85383-006-2


Set in 12/15pt Palatino
Typeset in Great Britain by AKM Associates (UK) Ltd, Southall,
London


Earthscan Publications Limited is a wholly owned, but editorially
independent, subsidiary of the International Institute for Environment
and Development (IIED).

# Contents

# Addresses of Authors

**Andrew Millington**
Department of Geography
University of Reading
Whiteknights
Reading
England, RG6 2AB

**John Townshend**
NERC Unit for Thematic Information Systems
Department of Geography
University of Reading
Whiteknights
Reading
England, RG6 2AB

**Pam Kennedy**
Department of Geography
University of Reading
Whiteknights
Reading
England, RG6 2AB

**Richard Saull**
Department of Geography
University of Reading
Whiteknights
Reading
England, RG6 2AB

**Steve Prince**
Earth Resources Branch
Code 623
NASA/Goddard Space Flight Center
Greenbelt
MD20771, USA

**Robert Madams**
World Conservation Monitoring Centre
219c Huntingdon Road
Cambridge
England, CB3 0DL

# Foreword

This volume emerged from work commissioned by the Technical and Administrative Unit (TAU) of the Southern African Development Co-ordination Conference (SADCC) Energy Sector, to explore ways of tackling a growing energy problem within the region. Wood is the people's fuel and, in a number of areas, people are finding it more and more difficult to obtain the necessary supplies. This book provides the first SADCC regional biomass assessment using remote sensing and its findings should help all those concerned with energy and development, both within and outside of the SADCC region.

The main overall conclusion of the wider study (of which this volume forms a part) carried out by the ETC (Consultants for Development) Foundation, is that the best way to ensure future fuelwood supplies, and simultaneously to prevent environmental degradation, is to improve the management of woody biomass within existing production systems based upon the innovations and responses already occurring amongst smallholder farmers. To put this into operation at national and regional levels, a new series of relationships have to be developed between those ministries responsible for energy, forestry, rural development, environment and agriculture. These "new relationships" will be the basis for enhancing future wood supplies. Most importantly, these new relationships can be used to develop research and extension networks that support woody biomass production by local farmers. This, of course, requires that fuelwood strategies must go beyond energy and forestry projects and be incorporated into as many other projects as possible.

Many people have contributed to the success of the SADCC Fuelwood Study, in particular, a large number of the region's own experts from various fields. Our thanks are due to all those who contributed their considerable expertise to the success of the project. The study was jointly financed by the Netherlands Government and the European Economic Community to whom thanks are also due.


*April 1988*                                                    *Jorge Tavares de Carvalho Simoes*
                                                               *Technical and Administrative Unit*
                                                                      *Regional Co-ordinator*
                                                                      *SADCC Energy Sector*

# Preface: The Fuelwood Problem in the SADCC Region

This book emerged as part of the output of the SADCC Energy Development Fuelwood Study which was commissioned by the SADCC Energy Sector based in Angola. It represents a first attempt at estimating stocks and yields for the region using remote sensing techniques in combination with secondary data; and provides a system of biomass classification for all nine SADCC member states. The strength of such an assessment is that it furnishes a broad situational analysis. The value of such a work will be appreciated over the long term when further exercises are undertaken at periodic intervals to facilitate comparative analysis of the changing biomass situation within the region. As more detailed analyses are undertaken of individual countries using different remote sensing techniques, a more complete picture will emerge of biomass supply.

*Biomass Assessment* addresses one particular aspect of fuelwood planning, namely existing supplies of woody-biomass material. To develop energy planning, four broad sets of data are required. These are:

   i)   data on commercial supply,
   ii)  data on commercial demand,
   iii) data on non-commercial supply, *and*
   iv)  data on non-commercial demand.

Data on commercial supply are relatively easy to obtain at national level from relevant government offices. Data on commercial demand can be derived from sales at energy consumption points and, although such data are occasionally commercially restricted, they are usually available to match the commercial supply figures. Data on non-commercial demand (i.e., demand for traditional fuels that are frequently beyond the marketplace) are obtainable through mass survey techniques and, while not as easily available as commercial data, can be generated.

The area for which there is least data is that of non-commercial supply. There are several reasons for this. First, existing forestry conventions only deal with forest areas, but the major source of fuelwood is trees outside the forest. Secondly, where forest inventories do exist, mensuration tables are drawn up on a volume not a weight basis, reflecting forest industries concerned with timber and pole production rather than energy content. Thirdly, even when mensuration tables exist and can be converted from volume to weight equivalents, they contain only minimal documentation of total woody biomass productivity, including branches and roots. Finally, few records are made of the woody scrub undergrowth in such forest areas. Needless to say, if this information is largely unavailable for the forest areas themselves, it is totally unavailable for the broad area of trees outside the forest, including natural woodland.

Existing forest inventories, however, must not be undervalued as they are usually the single most important source of estimates of woody biomass production. Coupled with a range of ecological and botanical site studies, they provide the broad base of secondary material on which any assessment of woody biomass is built. In order to interpret the quality of such data across the SADCC region, it is necessary to build the assessment around a common methodology. Remote sensing, using one data source, allows the production of a uniform assessment across the region which can be checked against this secondary data and verified in the field. This book describes such an exercise which has produced the first maps and numerical estimates of woody-biomass stock and yield throughout southern Africa.

Such an exercise is only a beginning. The process of drawing maps based on a variety of data sources is essentially an iterative process in which accuracy is dependent on the scale of mapping and the ability to verify results on the ground.

*Biomass Assessment* provides an interesting insight into the fuelwood problem. In terms of national woody biomass availability, there would not seem to be a situational crisis, but these national aggregate statistics mask real problem sites, especially in semi-arid areas and around larger cities.

This broad situational analysis provides a starting point. It does not, however, enable us to pinpoint specific problem areas, nor does it tell us about the real nature of the problem. For this, a more complete and disaggregated analysis of particular areas would be needed, one that would examine the following: population distribution patterns and their dynamics, which profoundly affect levels of demand; patterns of land tenure as an aspect of the particular problems given social groups encounter in obtaining fuelwood supplies; the economic activities being pursued in any specific area, which may greatly affect fuelwood availability (i.e., land clearance for agricultural production or tobacco production, which places a particularly heavy demand on woody biomass for curing); the need of urban centres for a great deal of timber, for construction as well as firewood; and the general structure of woody-biomass resource management and use, the starting point for any successful intervention strategy.[1]

To intervene successfully entails a process of focusing in, from the situational analysis outlined in this book to particular problem sites. Getting closer to the problems on the ground, however, requires knowing much more about the wider role that woody biomass plays in the local community.

**A NEW APPROACH TO THE FUELWOOD PROBLEM**

The findings of the overall study are outlined in *The Fuelwood Trap: A Study of the SADCC Region*.[2] Here, we provide no more than a synthesis of the arguments. Whilst *Biomass Assessment* gives a broad picture of woody biomass supply within the region, the scale is such that it can only pick up general categories over large areas. For example, whilst the slopes of Mount Kilimanjaro may be tree-covered, the peak is certainly not! The scale of this study is such that it is difficult to pick up this level of detail.

We know that in rural areas people rely mainly on fuelwood gathered outside of the forests, from trees and bushes scattered in and around the farm. For this and other reasons, previously outlined, particular problem areas need detailed investigation.

Traditional interventions have been frequently unsuccessful because they have not properly understood the problem. Assessments of biomass supply have been made and these have been compared with projected demand. If a shortfall exists between the calculated sustainable yield and the growing demand, then the "gap" between the two is used as a measure of the problem. This is faulty reasoning because first, a lot of wood is derived from trees outside the forest, and these are not usually included in calculations of supply availability; and secondly, there is an order of preference in the types of woody biomass given domestic use. When the preferred tree species are no longer available, shrubs and agricultural residues can be used instead. Hence, the dire projectons which emphasize that no wood will be available by a given year if existing supply and demand trends continue, are often not true in practice. Yet this does not mean that people have reliable supplies of fuelwood. The problem of diminishing fuelwood resources is real and it is growing.

The solutions proposed have been determined too often by policy conclusions derived from this "gap" analysis. The possible answers to the problem of the projected gap between available supply and projected demand have been twofold and predictable: increase supply and reduce demand. The solutions seized upon in the 1980s were based upon existing institutional capabilities and the promise of new technologies. Increasing supply meant using professional forestry skills to produce solutions to the fuelwood problem – what else but fuelwood trees? This was immediately translated into peri-urban plantations to meet the growing urban fuelwood demand and communal woodlots for the rural population.

This strategy for supply enhancement ignored the simple economics of the fuelwood market. Commercialized fuelwood (a term that includes both charcoal and firewood) is mainly an urban phenomenon. In the rural areas domestic energy supplies are a "freely" available good. The notion of a "freely" available good needs to be qualified; everthing has a cost. In this case it is overwhelmingly increased labour burden on women.[3] As the accessibility of fuelwood has diminished, women and children have spent a greater amount of time in fuelwood collection. Despite this, growing fuelwood could never compete economically with wood available without production costs in the rural areas. Neither could it compete with the potential of using land for agricultural production. Similarly, proposed solutions for conserving fuelwood by spreading the use of improved stoves, could not work in rural areas where a free alternative, the three-stone fire, is readily available.

The best place to start looking for solutions to the fuelwood problem is in the day-to-day life and local environment of the people. Woody biomass, in the form of trees and shrubs, is part of the integrated production system of peasant farmers. Trees provide many of the necessities of life. Woody biomass provides important dry season fodder, particularly in arid and semi-arid lands. Construction poles, furniture, tools, medicines, cosmetics, fruit, habitat for game, mulch to enrich the soil, protection against wind and water erosion, environmental protection, ornamentation, fruit, and spiritual sustenance; all these and much more besides are derived from trees.

The key to success lies in encouraging farmers to grow more trees themselves. They will not do this if fuelwood is to be the only benefit but they will be encouraged if trees and tree products can be shown to provide a solution, or partial solution, to their more urgent needs. One reason why communal woodlots failed is because they did not address priority needs and there was uncertainty over the distribution of benefits.

Intercropping can improve soil fertility and yields, and relieve the farmer of the heavy costs of fertilizers; fodder trees can help sustain livestock; in many areas there is a demand for construction poles and these can be a viable cash crop; a tree crop can be an important risk-minimization strategy to provide income in years of bad harvest, or to meet expenses such as medical treatment. These types of end-product or result can act as an incentive to grow more trees but fuelwood is always a residue product. Hence the fuelwood problem is best tackled *indirectly*.

Instead of relying upon expensive forestry department nurseries to provide seedlings, local farmers, schools and community organizations can be encouraged to develop their own nurseries. The Rural Afforestation Programme in Zimbabwe, for example, has shifted its emphasis in this way and seedlings can be produced at a tenth of the previous cost as a result.

By encouraging local farmers to improve their land-use practices, tree components can be integrated in a positive way to improve production systems. For instance, trees cultivated in and around the farm, reduce the burden on women for fuelwood collection . Such practices make a significant contribution to sustainable development strategies and are a further step forward in improving the intensification of agriculture, which is made necessary as land scarcity no longer permits traditional patterns of shifting cultivation to be maintained.

Another important aspect of this new approach to the fuelwood problem is that it recognizes the value of the farmers' existing tree-management practices. Understanding the knowledge base that is already there makes it easier to encourage the sharing of improved techniques, which may only exist in pockets of a community. This resource is not being currently harnessed although it represents a powerful weapon in the battle for rural development. It is easier to encourage improvements by starting with the knowledge base already existing within the local community, than to "parachute" in new technologies and ideas from above. A new approach has to avoid "top-down" solutions and concentrate instead upon a more democratic and participatory approach.

This approach is being adopted in an increasing number of the sectoral activities within the SADCC. For example, a recent publication of the SADCC Soil and Water Conservation and Land Utilization Programme has stated this explicitly and deserves quoting in full:

> The conclusions drawn from the history of conservation in the region, and from our experiences as the Coordination Unit, clearly point to the kind of projects and programmes that are aimed at improving the living conditions of the population and having the farmers themselves as the main social actors in conservation. Conservation *per se* should not be an isolated activity but done in connection with others: the integration of conservation into the farming system, and therefore the adoption of practices and techniques adapted to both the natural environment and farmers' modes of production are a must. In many cases only limited additional inputs are required, the most valuable resource being the farmers themselves.[4]

Such an approach will require institutional changes. Foresters must no longer hide in the forests and plantations but come out into the community. Extension activities within the farming community must be developed and an integration of traditional and scientific knowledge be achieved. This will require new training and a commitment to extend forestry activities. Ways need to be found to integrate a tree component into non-tree-specific projects. This new approach represents an exciting development opportunity which must be seized. Where this is being implemented within the SADCC region, there is already evidence of success. The challenge now is to channel resources towards putting this new approach into practice on a significant scale.

The authors of *Biomass Assessment* would like to take this opportunity to thank those people and organizations who provided much help and information during the image interpretation phase, especially International Forest Science Consultancy, Oxford Forestry Institute, Land Resources Development Centre, David Hall and Andrew Barnett. During the preparation of the various reports, help was provided by Sheila Dance, Kevin White, Sam Mutiso, Peter Kent and Jon Styles. Finally, we wish to thank C.J. Tucker (NASA/GSFC) for the provision of AVHRR data.

*Barry Munslow, Phil O'Keefe,*
*John Soussan, Adriaan Ferf*
*and Yemi Katerere*

*NOTES*
1. For a detailed country example (Malawi) of what this process entails, see B. Munslow, Y. Katerere, A. Ferf and P. O'Keefe, *The Fuelwood Trap: A Study of the SADCC Region* (London: Earthscan, 1988), Chapter 6.
2. Ibid.
3. See I. Dankelman and J. Davidson, *Women and Environment in the Third World* (London: Earthscan, 1988).
4. *Splash* (newsletter for the SADCC Soil and Water Conservation and Land Utilization Programme), Vol. 4, No. 1, 1988, p.3.

# 1. Introduction and Methodology

Woody-biomass resource assessments have rarely been undertaken on a regional scale for the purpose of energy planning. Where such assessments have been attempted, they have relied on secondary data derived from a variety of sources. This book illustrates the numerous problems involved in this type of survey and shows how they may be partially overcome by the integration of remotely sensed data, primary data collection, and secondary data.

Prior to this study, woody-biomass assessments for energy planning in the Southern African Development Co-ordination Conference (SADCC) region had been based on national forestry statistics and, to a much lesser extent, on ecological studies. The former are generally unreliable for such assessments due to the following reasons:

i) The level of resource assessment varies considerably between countries due to differentials in departmental funding, size of country, effectiveness of the forestry departments, and the techniques used. Therefore, within the SADCC region as a whole, estimates were inconsistent and regional assessments using a standard technique were a necessity.

ii) As most assessments had been made by foresters, they reflected the foresters' concern with commercial timber production rather than fuelwood. A change of emphasis was needed.

iii) Population growth and redistribution, changing land-use patterns, increased commercial timber exploitation, and the increased demands for indigenous fuels meant that previous maps and estimates of woody biomass had quickly become outdated. Up-to-date assessments were, and still are, required for current energy-planning initiatives.

iv) Ecological studies, which have the potential to provide more useful, quantitative information on total biomass, are relatively rare in this area.

In summary it can be said that the current levels of biomass resource potential (i.e., existing stock and annual productivity) in SADCC member states were not accurately known. Up-to-date ecological assessments made using a standardized methodology were urgently required.

It is essential in both regional and national energy budgeting that accurate, up-to-date information is obtained. Given the poor quality of the existing material, the only feasible method for providing this kind of information for a large region (such as SADCC) over a short period was to use remotely-sensed data. This type of data had already been used to a certain extent at the country level within the SADCC region, for Botswana, Mozambique and Tanzania. However, some of these studies were out of date by the time this study began and they had generally failed to provide the high levels of information that can be gained when using remotely-sensed data. This was due to the fact that many studies had not had access to digitally processed data, only photographic products. This can provide more information from remotely-sensed imagery than the photographic (optical) products of computer-aided techniques.

Ideally, biomass resource assessment is undertaken using remotely-sensed data combined with ground verification and calibration. In this project, biomass classes identified from remotely sensed data were verified initially with the use of secondary data sources such as natural resource reports and maps. This sometimes hampered effective analysis because of problems related to data quality, differences in levels of ground data between biomass classes (in some instances there was no data at all), interpretation of different types of data, and ground location. Therefore, the level of interpretation was variable and subsequent field verification of key areas was undertaken to check ground interpretations and collect growing stock data. This fieldwork was undertaken in Malawi, Swaziland and Zimbabwe during August 1986.

**REMOTE SENSING**

Satellite remote sensing is the technique of acquiring data (imagery) about the earth surfaces and atmosphere from satellites. Sensors on board satellites provide data at a synoptic scale from a single data-take, thereby making the analysis of remotely-sensed data a cost-effective tool for resource assessment over large areas. A further advantage of satellite data is that repeat imagery for any scene can usually be acquired. The repeat period between data-takes varies from relatively short periods of less than 24 hours on meteorological satellites to between 16 and 26 days on earth resource satellites. These figures represent the shortest repeat times and, due to a variety of technical and logistical reasons, when using earth resource satellites the repeat periods can be much longer. The capability to provide time-sequential data is, however, of great importance in both resource assessment and monitoring changes in the resource base.

Sensors for the acquisition of remotely-sensed data are found on various types of satellites which can be divided into two groups:

i)   earth resources satellites (e.g., the American Landsat series and the French SPOT satellite); *and*
ii)  meteorological satellites (e.g., the NOAA series and METEOSAT).

Data are provided to the user either as photographic (optical) products or computer compatible tapes (CCTs). The latter format is more useful for resource assessment as the data can be digitally processed and the maximum amount of information can be extracted from an image. This, however, can only be carried out effectively using computerized digital image-processing systems.

The data used in this work were acquired by the Advanced Very High Resolution Radiometer (AVHRR). This sensor is carried on board the American NOAA meteorological satellites, in this case NOAA-7, and the data were supplied on CCTs from NASA. The salient characteristics of this satellite and sensor combination have been detailed by Kidwell (1984) and Justice (1986) and its main characteristics are outlined in Table 1.1.

**Table 1.1** CHARACTERISTICS AND STATUS OF NOAA WEATHER SATELLITES AND THE AVHRR SENSOR

*Satellites:*

| | | | | |
|---|---|---|---|---|
| TIROS–N | Launched: | Oct. 1978 | Current status[1]: | Non-Operational |
| NOAA 6 | Launched: | Jun. 1979 | Current status[1]: | Non-Operational |
| NOAA 7 | Launched: | Jun. 1971 | Current status[1]: | Non-Operational |
| NOAA 8 | Launched: | Mar. 1983 | Current status[1]: | Non-Operational |
| NOAA 9 | Launched: | Dec. 1984 | Current status[1]: | Non-Operational |
| NOAA 10 | Launched: | Oct. 1986 | Current status[1]: | Operational |
| NOAA 11 | Launched: | Sep. 1988 | Current status[1]: | Operational |

*AVHRR Sensor:*

| | |
|---|---|
| Coverage cycle: | 9 days |
| Ground coverage: | 2,700 km |
| Orbital height: | 833 km |
| Orbital period: | 102 min |
| Ground resolution: | 1.1 km to 2.4 x 6.9 km$^2$ |
| Spectral channels: | 5 – Ch.1 580 – 680 nm |
| | Ch.2 725 – 1,100 nm |
| | Ch.3 3,550 – 3,930 nm |
| | Ch.4 10,300 – 11,300 nm |
| | Ch.5 11,500 – 12,500 nm |

[1] as of November 1988
[2] dependent on angle of view

**REGIONAL BIOMASS ASSESSMENT**

**Role of regional biomass assessment in the SADCC Fuelwood Project**

Regional biomass assessment was a fundamental aspect of the entire project as it provided the main supply parameters for energy-balance modelling and budgeting in the SADCC region and of the individual member states. Information relating to the areal distribution, quantity, and quality of fuelwood was provided at three levels:

    i)   the SADCC region;
    ii)  individual member states; *and*
    iii) individual administrative units in each member state.

This information enabled both the quantity and quality of woody biomass and its areal distribution to be assessed. It involved two parts: first, the identification and mapping of biomass classes; and secondly, the construction of a woody biomass database of biomass class areas, growing stocks, and mean annual increments (MAIs).

**Identification and mapping of biomass classes**

This component of the work was undertaken in three phases:

    i)   *Phase I:*  Inspection and preprocessing of AVHRR data;

    ii)  *Phase II:*  Initial image interpretation and mapping of vegetation classes, derivation of temporal data plots, and training statistics; *and*

    iii) *Phase III:* Automatic classification of biomass classes.

**AVHRR Data**

Data are acquired for any part of the earth's surface by the AVHRR sensor on board each NOAA satellite. These data are aimed primarily at the meteorological community and are taken every 12 hours. Therefore, every alternate data-take is at night and is of limited use in the context of biomass assessment. However, at the equator, the daylight pass for NOAA 7 and 9 satellites was 14.30 hours (local time) and on NOAA 6,8, and 10, it was 07.30 hours (local time). Because of the short repeat-period between data-takes, there is a very high probability that in any weekly period, a ground resolution element (the smallest area on the ground that can be resolved on the imagery) will be cloud-free during data acquisition and the reflectance of the land-cover sensed rather than that of the cloud-top.

Because of this ability to acquire information about the land surface, the potential of AVHRR data in a wide variety of ecologically-based studies has become apparent in recent years (Gatlin *et al.*, 1983; Norwine & Greegor, 1983; Schneider *et al.*, 1981; Tarpley et al., 1984; Townshend & Tucker, 1984; Tucker *et al.*, 1983, 1984a, 1984b, 1985a, 1985b). The configuration of the AVHRR sensors, particularly the wavelengths of the sensors in Channels 1 and 2 (red and near infra-red respectively), means that when cloud-free situations occur, useful ecological information can be obtained from within the ground resolution elements. The data from these two channels are combined for each ground resolution element and an index of vegetation status is calculated. This is called the Normalized Difference Vegetation Index (NDVI) and it is calculated by the following equation.

$$NDVI = \frac{Channel\,2 - Channel\,1}{Channel\,2 + Channel\,1}$$

The nominal resolution of the AVHRR sensor is 1.1 km, although a variety of operational raw data products from this base resolution up to 25 km have been produced by NOAA (Kidwell, 1984). NOAA have for certain years produced Global Vegetation Index (GVI) data on a weekly basis; being a sub-sampled temporally composited NDVI product that covers most parts of the globe on a polar stereographic projection with a nominal resolution of 20 km.

Although such data have been used for land cover discrimination of large areas (e.g. Tucker *et al.*, 1985a, and Townshend *et al.*, 1987) for this study it was felt that this resolution was too coarse and 8 km GAC (Global Area Coverage) data produced at NASA and available for Africa was used in most countries. In Lesotho, Swaziland and the adjacent areas of southern Mozambique, 4 km data was used. Most of the region is out of range of regular ground-based 1.1 km resolution data collection so a special request has to be made for this data to be collected.

The use of vegetation indices for vegetation mapping of Botswanan rangelands have been criticized by Ringrose and Matheson (1988). A more recent study in The Gambia (Laangas, 1987) questioned the use of AVHRR data for forest monitoring. Despite these two studies, the majority of biomass resource assessment and monitoring studies using AVHRR data have vividly illustrated the potential of these data.

Biomass classes have been identified and mapped for the entire SADCC region using GAC data. Whilst the use of 8 km data has provided good information for mapping woody biomass resources over the entire SADCC region and most individual member states, it has not provided data of sufficient spatial resolution to accurately map biomass classes in Lesotho and Swaziland. Finer resolution Global Area Coverage NDVI data, with a 4 km resolution, was used for biomass class mapping in these two countries.

## Image Processing

### Phase 1

All image processing was undertaken on 1²S 500 and 102 image-processing systems hosted by Vax 11/750 and HP 3000 computers respectively. Images were initially evaluated on the monitor, and hard-copy imagery for further interpretation and publication was taken using a Dunn Camera.

NASA provided NDVI data for 1984 which was used in this work: it was taken at two-weekly intervals for 4 km resolution and at monthly intervals for 8 km resolution. The monthly data were initially inspected for defects and two types were found:

i) pixels without monthly NDVI values; *and*
ii) misregistration between monthly data sets.

Pixels, a contraction of picture elements, are the equivalent of the ground resolution elements in the processed data. Pixels without data resulted either from sensor, transmission, or processing errors, or due to cloud-screening using thermal bands. In many cases, those with missing data were single pixels and were eliminated from the data by passing a 3 x 3 pixel median filter over it. This is a matrix of the 9 pixel values in a 3 x 3 square. It enables empty central pixels to be filled with an NDVI value based on the 8 adjacent pixel NDVI values.

Larger areas with no data (also due to cloud screening) were also encountered, which affected mainly Angola, Tanzania and Lesotho. None of these areas were present in all monthly data sets and, by elimination of the months with the largest "holes", the data were readily processed and interpreted. In the final biomass-class maps, the "holes" were eliminated by reference to surrounding biomass classes and ecological trends.

The data for December 1984 were misregistered with the data for the other months by 5–7 pixels. This misregistration proved difficult to successfully re-register and the December data were discarded.

*Phase II*

Image processing of monthly NDVI data provided various types of imagery for the initial identification and mapping of vegetation types. The types of imagery used in the Phase II interpretation were:

i) individual monthly NDVI images prepared for all months except December. These were supplied directly from NASA and were colour-coded on the basis of the digital numbers representing the NDVI values.
ii) an integrated NDVI image created by a pixel-by-pixel addition of all monthly NDVI values for corresponding pixels. The total NDVI values for each pixel were then divided by 11, that is, the number of months used. This

image provides a good indication of the amount and variation in annual vegetation productivity.

iii) difference images between two monthly NDVI values created by subtracting one set of monthly NDVI data from another on a pixel-by-pixel basis for corresponding pixels. The months used to create the difference images were chosen on the basis of the monthly NDVI values. Difference images were found to be particularly useful if imagery from the low NDVI period (August to October, across the SADCC region as a whole) was subtracted from the high NDVI period (January to April). This provided a good index of vegetation seasonality. The most useful image in this respect was February NDVI–September NDVI. Another difference image, May NDVI–June NDVI, was found to be useful in assessing the rates of vegetation dieback between the wet and dry seasons for different vegetation types.

iv) classified imagery produced using monthly NDVI data for February, May, July, and September. These months were chosen as representative of the high and low NDVI periods in the region (February and September respectively), and the intermediate periods, when NDVI values decrease at varying rates depending on vegetation types. These data were then clustered using a nearest-neighbour algorithm in which pixels with a statistically similar range of values over the four months were clustered together into the same class.

v) principal component imagery. Principal components analysis was carried out on NDVI data from all 11 months. The inverse values of the principal components were obtained by inverting the matrix of principal components obtained from the initial analysis. The inverted Principal Component 1 was found to be a good indicator of total annual productivity.

vi) false colour composite (FCC) imagery made using the above data. FCC images are useful as they allow the simultaneous interpretation of three derived images. They are created by assigning the three different data sets to the three colour guns (red, green and blue) of the colour monitor on the image-processing system – the so-called "false colours". Variations in colour in the final image can then be used for interpretive purposes. Three FCCs provided useful information in this work, they were: (a) integrated NDVI, February NDVI – September NDVI, May NDVI–June NDVI; (b) Principal Component 4, Inverted Principal Component 1, Principal Component 2; (c) Principal Component 3, Principal Component 4, Principal Component 5.

All of the above images were evaluated visually on the monitor and then photographed. The photographs were used to identify and map the vegetation types on a 1:5,000,000 scale base map. Identification of vegetation types was based on the following criteria:

i) vegetation phenology obtained from graphs of monthly NDVI plots taken from selected sites within the SADCC region and the seasonality images;

ii) monthly and annual vegetation productivity based on monthly and annual NDVI imagery, and principal component imagery;

iii) secondary information based on ecological and forestry research; *and*

iv) other environmental information, particularly geological data, soils, topographic maps, and climatic statistics.

Nineteen broad regional vegetation classes were identified in the SADCC region using the 8 km data (that is, excluding Lesotho and Swaziland). These were related to different types of vegetation communities (forest, woodland, shrubland, bushland, grassland, or desert). Their distribution does not exactly correspond to any one previous botanical or ecological study of the region, but they are broadly comparable to most regional maps of southern and central African vegetation.

**Phase III**

The vegetation types that were identified and mapped throughout the SADCC region (Phase II) were used as the basis for classifying the region into biomass classes. Biomass classes are defined as areas with a vegetation type, or types, which are comparable in terms of overall biomass, productivity, and seasonality. They may, however, combine more than one type of ecological structure (such as forests and woodlands) and many floristic units.

The biomass classes were identified using a supervised classification algorithm. Supervised classification is a particularly powerful image-processing technique in which the values of each pixel are compared to various sets of pixel values (training statistics) from a number of known areas (training sites). The individual pixels are assigned to the most suitable class based on the training-site statistics. Different algorithms can be used in supervised classifications, but in the SADCC work a maximum likelihood classifier was used. Training sites were identified on the basis of Phase II interpretations. These were located in areas where the 19 vegetation classes had their maximum areal extent. Training statistics for all the vegetation types were generated and the data were classified on the basis of these. This produced an automated classification of 19 biomass classes throughout the SADCC region. The NDVI values for the 11 months, and the first three principal components (cf., Phase II) were classified separately and examined.

Interpretation and evaluation of the two images showed that the classification of biomass classes based on the 11 monthly NDVI values were most meaningful. These biomass classes were then reinterpreted using the same criteria as in Phase II and the 1:5,000,000 map generated in Phase II. Three of the biomass classes (montane forest, wet miombo woodland and cloud forest) were found to be phenologically indistinct throughout the region and were subsequently merged, leaving a total of 17 biomass classes.

Biomass class maps for Lesotho and Swaziland were generated using a similar procedure to that already outlined above but using 4 km spatial resolution GAC data. Because of the strong floristic links between Swaziland and the areas of Mozambique south of the Limpopo River, the latter area was reinterpreted using 4 km data.

The automatically classified data provided the final product in terms of both regional and national biomass class maps. These maps are interpreted individually for the entire SADCC region (see Chapter 2) and in the detail for each country (see Chapters 3-11). Biomass assessments are produced as both maps and tables. The final maps for each country and the SADCC region are based on computer-generated digital maps of the biomass classes. In the case of Malawi, Swaziland and Zimbabwe these have been field-checked; the maps for the other countries have been revised by the appropriate national Energy and Forestry Departments. In addition, statistical summaries of the growing stock and productivity of the different biomass classes are provided for the SADCC region, individual member states, and the administrative units in each country.

**Database for biomass
estimation**

Woody biomass has not been mapped directly by the methods outlined above. Instead, the technique has used the temporal change in spectral response to define land-cover types which are reasonably internally homogenous in terms of biomass. Primary and secondary data on the woody-biomass component (trees and shrubs) of these classes has then been built up into a woody-biomass database.

Ideally, the availability of woody biomass in each SADCC country should be evaluated using detailed information gathered at a local or district level. However, such information does not exist and it is possible only to obtain generalized and often extrapolated data. In the light of this, biomass information was sifted from a scanty literature source in an attempt to provide data for the biomass classes previously identified from the image-processing and interpretation phases (Table 1.2).

Production and biomass are measured in various units such as volume, carbon content, or energy content. However, the most practical and commonly available data for the main vegetation types in southern Africa were those of mass of dry material per unit area for biomass, and mass per unit area per year for production. As a consequence, all figures in the following assessment are given in tonnes and tonnes per year, for biomass and production respectively, for above-ground parts only.

It should be emphasized that a certain degree of selectivity has been exercised in the incorporation of data from available literature. In a few cases, data have been included from sources that do not state sufficient details of method for full interpretation, but these data have only been used where no other data are available for the biomass class concerned.

These data (see Chapters 3–11) include theoretical and empirical estimates extracted from a few detailed localized projects and general countrywide studies. In cases where a number of studies provide multiple estimates for a single biomass-class, average figures are quoted. All these figures have been extended to provide estimates for the specific biomass classes identified within each SADCC country (see Tables 1.3 and 1.4). These data, in turn, have been adopted to estimate total biomass when studied in association with the calculated area of each biomass class.

In order to appreciate the validity, utility, and limitations of these figures, it is essential that the general problems associated with biomass estimation and the difficulties encountered in obtaining

**Table 1.2** COMPILATION OF WOODY BIOMASS SUPPLY DATABASE

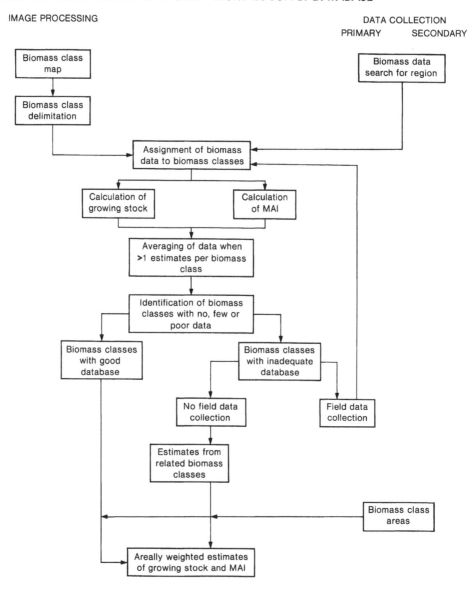

directly comparable data are fully understood. The most important points are:

i)   the need to distinguish between timber for manufacturing and construction, and that for fuelwood. This illustrates one of the greatest problems in estimating fuelwood yields from natural forests and woodlands since it is commonly only the "wood residue" (i.e., branches and twigs) which are collected and used for fuelwood and not the tree trunks. Unfortunately, few studies recognize this distinction because they are either forest inventories for estimating timber production, or ecological studies concerned with all the vegetation components. Accurate estimates for fuelwood resources often fall beyond the scope or requirements of the commercial wood estimates, and it is often difficult to extract the relevant biomass information.

ii)   the danger of over-estimating fuelwood yields based on biomass estimates for particular biomass classes when certain factors are not taken into consideration (for example, accessibility, selective preferences for certain species, and the reservation of forests and woodlands as Government Reserved Areas).

iii)   the fact that the majority of total-biomass estimates are for a single period in time and frequently omit to provide an indication of seasonal variations and long-term changes caused by, for example, deforestation and climatic change. However, the seasonal variations can be in biomass production provided by plotting NDVI values for each month.

iv)   other recognized sources of discrepancy often give rise to gross spatial and temporal variations in biomass. These are: (a) the method of derivation (e.g., establishing a relationship between tree/shrub biomass and canopy-stem diameter); (b) the ratio of tree biomass to shrub biomass per unit ground area; (c) the morphology of the species (in terms of stem wood, branch wood, leaf and twig biomass components); and (d) the vertical and lateral distribution of biomass depending on the heterogeneity of the vegetation community.

The combined effect of all these factors is that the general availability of published data is extremely variable, both for each country and for the different vegetation communities. Consequently, biomass estimates for the same biomass class vary significantly both between countries and within any one country.

As a consequence of the problems outlined above, the database for estimating standing woody biomass (Growing Stock) and woody biomass productivity or Mean Annual Increment (MAI) for the different ecological environments within the SADCC region is deficient in some areas and highly variable in terms of derivation and accuracy. Therefore, it is essential to emphasize that the tabulated data for each country (see Chapters 3–11) should not be treated as rigorously sampled scientifically accurate biomass data. The data should be seen as indicative of the level of growing stock and productivity. The data from each class may not be strictly comparable but the relative variations are important. In any circumstances the data should only be used with the caveats outlined above and treated with a certain amount of caution.

Caution should also be exercised when accepting the validity of drawing direct and indirect relationships between data extracted from image processing techniques and from a literature search. The difficulties of making links reinforces the need for fieldwork in selected areas of high and low fuelwood potential, in order both to verify the results obtained from this study and to increase the accuracy of biomass estimates for a small sample of the identified biomass classes. Such fieldwork was undertaken in August 1986, in the main biomass classes found in Malawi and Swaziland. Sample plots were selected on the basis of random sampling in the biomass-class maps of each country. Sample sizes varied from 100–400 m² and observations were made of the following:

   i)   tree dbh (trunk diameter at 1.30 m above the ground);
   ii)  tree height;
   iii) tree species;
   iv)  shrub (<1.5 m high) counts; *and*
   v)   evidence of felling, coppicing and pollarding.

The biomass data from these sites were calculated using the equations provided for Zambian miombo woodlands (Stromgaard, 1985a) for Malawi, and Botswanan *Acacia* woodlands (ERL, 1985) for Swaziland. These data were then incorporated into the biomass database (see Tables 1.2 and 1.4).

**Table 1.3**  BIOMASS CLASS EQUIVALENTS FOR THE SADCC REGION

*Class A*
Angola:        Transitional Rain Forest/Miombo Woodland

*Class B*
Angola:        Dense, High Miombo Woodland
Botswana:      Riparian Forest
Mozambique:    Evergreen Miombo Woodland and Coastal Forest
Tanzania:      Wet Miombo Woodland
Zambia:        Dry Evergreen Forest
Zimbabwe:      Dense Savannah Woodland

*Class C*
Angola:        Dense, Medium-Height Miombo Woodland
Botswana:      Dense Woodland
Malawi:        Evergreen and Semi-Deciduous Forest and Woodland
Mozambique:    Wet Seasonal Forest and Woodland
Tanzania:      Wet Seasonal Miombo Woodland
Zambia:        Wet Miombo Woodland
Zimbabwe:      Open Savannah and *Baikiaea* Woodland/Montane Vegetation

*Class D*
Angola:        Seasonal Miombo Woodland and Wooded Savannah
Malawi:        Seasonal Open-Canopy Miombo Woodland
Mozambique:    Seasonal Miombo Woodland
Tanzania:      Dry Miombo Woodland
Zambia:        Seasonal Miombo Woodland
Zimbabwe:      Seasonal Savannah Woodland

*Class E*
Angola:        Dry Deciduous Savannah
Malawi:        Miombo Woodland with Tobacco Cultivation
Zambia:        Kalahari Woodland

*Class F*
Tanzania:      Cleared Miombo Woodland
Zambia:        Degraded Miombo Woodland

*Class G*
Mozambique:    Coastal Forest Mosaic
Tanzania:      Coastal Forest Mosaic

*Class H*
Angola:        Degraded Rain Forest and Miombo Woodland
Malawi:        Dry Open-Canopy Miombo Woodland and Cultivation Savannah

*Table 1.3 Continues*

*Table 1.3   Continued*

| | |
|---|---|
| Mozambique: | Dry, Miombo Woodland/Wood and Shrub Thicket |
| Tanzania: | Semi-Arid Steppe |
| Zambia: | Dry Miombo and Munga Woodland |
| Zambabwe: | Dry Savannah Woodland |

*Class I*

| | |
|---|---|
| Botswana: | Open Woodland |
| Malawi: | Mopane Woodland |
| Mozambique: | Mopane Woodland |
| Zambia: | Mopane Woodland |
| Zimbabwe: | Mopane Woodland and Escarpment Thicket |

*Class J*

| | |
|---|---|
| Angola: | Dry Inland Savannah/Dry Coastal Savannah/Arid Coastal Thicket |
| Malawi: | Swamp and Lake Vegetation/Tea and Coffee Cultivation |
| Mozambique: | Dry Savannah Woodland |
| Zambia: | Swamp and Lake Vegetation |
| Zimbabwe: | Intensive Commercial Arable Agriculture |

*Class K*

| | |
|---|---|
| Angola: | *Chanas da Borracha* Grassland/Degraded Dry Deciduous Savannah |
| Botswana: | Woodland and Bushland |
| Mozambique: | Degraded Agricultural Land |
| Swaziland: | Dense Bushland and Woodland |
| Zambia: | Scrub Woodland |
| Zimbabwe: | Dry Bushy Savannah |

*Class L*

| | |
|---|---|
| Botswana: | Bushland with Scrubby Woodland and Woody Shrubland |

*Class M*

| | |
|---|---|
| Botswana: | Shrubland and Bushy Shrubland |
| Swaziland: | Sparse Woodland and Bushland/Open Grassy Savannah |
| Zimbabwe: | Degraded Savannah |

*Class N*

| | |
|---|---|
| Angola: | Montane Grassland |
| Botswana: | Hill Shrubland and Woodland |
| Mozambique: | Dry Riparian Woodland |
| Tanzania: | Semi-arid Dry Steppe |
| Zimbabwe: | Wooded Grassland |

*Table continues*

*Table 1.3 Continued*

### Class O
| | |
|---|---|
| Angola: | Bushy Arid Shrubland |
| Botswana: | Fringing Palm Woodland |

### Class P
| | |
|---|---|
| Angola: | Desert Vegetation/Coastal Vegetation |
| Botswana: | Salt Pans |

### Class Q
| | |
|---|---|
| Mozambique: | Lubombo Hills Woodland |
| Swaziland: | Lubombo Hills Woodland |

### Class R
| | |
|---|---|
| Lesotho: | Escarpment Grassland with Scrub Woodland |

### Class S
| | |
|---|---|
| Lesotho: | Highveld and Riparian Grassland |

### Class T
| | |
|---|---|
| Lesotho: | Alpine and Sub-Alpine Grassland and Heathland |

### Class U
| | |
|---|---|
| Mozambique: | Lowland Sublittoral Forest and Bushland |

### Class V
| | |
|---|---|
| Mozambique: | Littoral Grassland |

### Class W
| | |
|---|---|
| Lesotho: | Escarpment and Riparian Woodland |
| Swaziland: | Highveld Forest |

### Class X
| | |
|---|---|
| Swaziland: | Dense Plantation Stands |

### Class Y
| | |
|---|---|
| Malawi: | Plantations (*Eucalyptus/Gmelina*) |
| Swaziland: | Wattle and *Eucalyptus* Plantations |

### Class Z
| | |
|---|---|
| Swaziland: | Irrigated Agriculture |

**Table 1.4** SADCC: DERIVATION OF GROWING STOCK AND MAI

| Biomass Class (Table 1.3) | Growing Stock (t/ha) | MAI (t/ha/yr) | Yield ratio (MAI/ Growing Stock x 100) | References |
|---|---|---|---|---|
| A | 71.22 | 2.25 | 3.16 | NO DATA |
| B | 71.22 | 2.25 | 3.16 | NO DATA |
| C | 71.22 | 2.25 | 3.16 | Guy, 1970<br>Guy, 1981<br>Persson, 1981<br>Stromgaard, 1985<br>Field Data, 1986 |
| D | 19.85 | 0.49 | 2.47 | Persson, 1981<br>Field Data, 1986 |
| E | 16.79 | 0.49 | 2.92 | Rushworth, 1975<br>Field Data, 1986 |
| F | 33.55 | 1.01 | 3.01 | Stromgaard, 1985 |
| G | 246.77 | 29.49 | 11.95 | Christensen, 1978<br>Golley et al., 1962 |
| H | 9.44 | 0.24 | 2.54 | Persson, 1981<br>Field Data, 1986 |
| I | 36.97 | 1.09 | 2.95 | Blair-Rains & MacKay, 1968<br>ERL, 1982<br>Guy, 1970, 1981<br>Kelly & Walker, 1976<br>Rutherford, 1975 |
| J | 11.8 | 0.42 | 3.55 | NO DATA |
| K | 23.36 | 0.83 | 3.55 | ERL, 1982<br>Guy, 1980<br>Mills, 1975<br>Field Data, 1986 |
| L | 14.22 | 0.60 | 4.22 | ERL, 1982<br>Field Data, 1986 |
| M | 7.11 | 0.30 | 4.22 | NO DATA |
| N | 11.45 | 0.55 | 4.80 | ERL, 1982<br>Persson, 1981 |
| O | 7.11 | 0.30 | 4.22 | NO DATA |
| P | 0 | 0 | 0 | NO DATA |
| Q | 75.92 | 2.65 | 3.49 | Field Data, 1986 |

*Table continues*

*Table 1.3 Continued*

| | | | | |
|---|---|---|---|---|
| R | 0.50 | 0.01 | 2.00 | Geobel *et al.*, 1986 |
| S | 0 | 0 | 0 | NO DATA |
| T | 0 | 0 | 0 | NO DATA |
| U | 9.44 | 0.24 | 2.54 | NO DATA |
| V | 0 | 0 | 0 | NO DATA |
| W | 280.84 | 2.97 | 1.06 | Field Data, 1986 |
| X | 39.50 | 5.16 | 13.06 | Keay & Turton, 1970<br>Keay *et al.*, 1980<br>McKee & Shoulder, 1974<br>Madgwick *et al.*,<br>    1977 a, b<br>Orman & Will, 1960<br>Turton & Keay, 1970<br>Whitaker *et al.*, 1975<br>Will, 1959<br>Will, 1964 |
| Y | 42.08 | 8.53 | 20.27 | Banks & Metelerkamp,<br>1969<br>Govt of Swaziland, 1981 |
| Z | 0 | 0 | 0 | NO DATA |

**Notes**

GS = growing stock

MAI = mean annual increment

*Class A*  Assumed equivalent to Class C

*Class B*  Assumed equivalent to Class C

*Class G*  Estimates refer to mangrove forests. Class G only contains these in part with other woodland classes on higher ground. Less productive GS/MAI have therefore been calculated for purposes of this work as (Class G x .1) + (Class B x .9) giving a GS of 88.78 t ha$^{-1}$ and an MAI of 4.98 t ha$^{-1}$ yr$^{-1}$.

*Class J*  Assumed as 50% of Class K. When J = Swamp and Lake Vegetation, Tea and Coffee Cultivation or Intensive Commercial Agriculture, then GS/MAI — 0.

*Class K*  When K = *Chanas da Borracha* grassland then GS/MAI = 0.

*Class M*  Assumed as 50% of Class L

*Class N*  When N = Montane Grassland then GS/MAI = 0

*Class O*  Assumed as 50% of Class L

*Class U*  Assumed as equivalent to Class H

*Class W*  GS/MAI for forest stands. As woodland occurs in very small plots, it is assumed to be never greater than 10% of pixel. Therefore GS = 280.84 x 0.1 = 28.08 and MAI = 2.97 x 0.1 = 0.30.

# 2. Overview of the SADCC Region

The purpose of this overview is to introduce the biomass classes in a broad context; to examine their distribution in the SADCC region; to highlight their woody-biomass supply potential; and to identify, on a regional basis, areas with low woody-biomass supply.

**BIOMASS CLASSES**

The biomass classes are defined by the seasonal variations in their growing patterns (phenology) and biomass characteristics. The phenological plots and biomass estimates were derived from the NDVI data in combination with secondary data. All vegetation indices exploit the variations in spectral characteristics of vegetation throughout the reflected spectrum. Vegetation canopies absorb radiation in red wavelengths for photosynthesis and reflect radiation in near infra-red wavelengths because of the structure of plant cell walls (Colwell, 1974). Ratios, or linear combinations, of measurements of reflected radiation in these wavelengths will therefore carry information concerning the status of vegetation.

The NDVI has been shown to be correlated with a variety of vegetation parameters ranging from biomass (Tucker *et al.*, 1985b) to atmospheric $CO_2$ (Tucker *et al.*, 1986). Detailed studies on the characteristics of the NDVI and its applications in Africa can be found in Justice (1986). In this study, the normally fractional value of the NDVI has been converted into a byte scale to make computer-processing easier. As the NDVI increases, there is greater vegetation activity. This can either be in the form of increased vegetation growth and leaf area index, and hence increased biomass, or increased photosynthetic activity in the canopy, which would lead to increased dry matter accumulation and more biomass. The minimum NDVI encountered is approximately 125. At this level there is, at most, a trace of vegetative activity. The maximum value encountered is in the region of 200, which indicates vigorous vegetative activity and correspondingly high levels of biomass accumulation.

As a consequence of the biomass classes being defined by growing stocks and productivity, they can cut across or amalgamate more familiar botanical mapping units based on floristic considerations. Their method of derivation is, however, more appropriate to the assessment of woody-biomass supply than botanical units because productivity and growing stock assessments can be inferred from them.

It should be noted that NDVI indicates more than just wood production. All plant material is considered, that is wood and leaves on trees, grasses, herbs, and crops. Consequently, wood production cannot be estimated directly from the data and reference must be made to ground reference data (see p. 11).

**Distribution of Biomass Classes**

The biomass classes in the SADCC region and their distribution are discussed below. More detailed descriptions are provided in the country profiles.

**Biomass Class A** is only found in northern Angola and represents the transition between the Rain Forest vegetation of the Zaire Basin and the Miombo Woodlands of the Central African plateaux. It is characterized by generally high levels of biomass productivity but there is a marked dry season lull.

**Biomass Class B** is mainly found in Angola, Mozambique, Tanzania, and Zambia. It represents evergreen closed canopy woodland, and dense woodland or forest with little seasonality and high levels of productivity throughout the year. Evergreen Miombo Woodland, other Evergreen Forests, and Riparian Woodlands also fall into this class.

**Biomass Class C** is found in a belt stretching from the Bie Plateau in Angola, across Zambia and Malawi, and into southern Tanzania and northern Mozambique. There is a further large area in Zimbabwe and central Mozambique. It represents two vegetation

21

types: Dense Wet Miombo Woodland with a slight seasonality and Montane Forests and Grasslands. The level of productivity is high, but there is a slight drop during the dry season that is not found in Evergreen Forests and Woodlands.

**Biomass Class D** represents the Miombo Woodland with a marked seasonality, having high productivity levels during the wet season and very low levels during the dry season. It is found scattered throughout the Miombo Woodland belt but its largest continuous areas are found in Angola, Mozambique and Zambia.

**Biomass Class E** mainly consists of drier woodlands than those of Classes B to D. Typical woodland types are those found on the Kalahari Sands in Angola, Botswana, Zambia and Zimbabwe. These are generally open-canopy woodlands with lower overall productivity levels than the wetter Miombo Woodland types, and a less marked seasonality. They include drier Miombo Woodland, Kalahari Woodland, and Dry Deciduous Savannah Woodland communities.

**Biomass Class F** represents areas where there has been extensive disturbance of the Miombo Woodlands and a wooded grassland community has established itself. The disturbance is usually due to shifting cultivation. It is found extensively in Tanzania and Zambia. In other countries the disturbance and degradation of Miombo Woodland leads to other vegetation types. Productivity is strongly seasonal and growing stocks are variable, but always less than in adjacent woodlands.

**Biomass Class G** is found along parts of the Indian Ocean coastal belt and it represents Coastal Forest. Productivity is high and varies little throughout the year. It includes Moist Coastal Forest, Mangrove Swamp Forest, and Estuarine Swamp Forest.

**Biomass Class H** is found extensively in the east of the SADCC region. In Tanzania it represents steppe bushland and thicket communities. However in Malawi, Mozambique, Zambia and Zimbabwe as well as other countries with smaller areas, it is indicative of dry, and sometimes degraded, Miombo Woodland. The wet season Productivity levels are generally high during the wet season but there is a long dry season when they are very low.

**Biomass Class I** represents very dry woodland or wooded savannah, and the class is dominated by Mopane Woodland. It is found in a belt stretching from northern Botswana northwards to north-east Zambia, and in an easterly direction to Malawi and Mozambique. Levels of productivity are generally moderate and there is a long period of moisture stress when production is limited.

**Biomass Class J** is found throughout the SADCC region. It represents a number of vegetation types all of which are characterized by moderate biomass levels (although the woody component is always low) and a marked, but not insignificant, seasonality. Swamp and Lake Vegetation in Malawi and Zambia is included in this class but more usually the class is indicative of intensive agricultural activity (e.g., in Angola, Malawi, Mozambique and Zimbabwe).

**Biomass Class K** represents areas of grassland (especially in Angola and Zambia) and (in Angola, Botswana, Mozambique, Swaziland and Zimbabwe) dry degraded scrubby savannah woodland. It occurs extensively in the drier parts of Mozambique and Zimbabwe, and the wetter parts of Botswana. This latter area is related to the areas in Swaziland. Productivity levels are low and relatively constant throughout the year.

**Biomass Class L** is restricted to Botswana where it indicates scrubby bushland in degraded areas. Productivity levels are low with a marked dry season lull.

**Biomass Class M** is found in the Kalahari Desert in Botswana, and related areas in Swaziland. In both countries it represents *Acacia*-dominated scrubby and shrubby vegetation. Productivity increases slightly during the wet season but is generally low throughout the year.

**Biomass Class N** occurs in scattered patches throughout the region. The main area is found in Tanzania where it represents a dry steppe environment. In other areas it represents scrubby and often degraded vegetation. The wet-season productivity levels are relatively high but fall off markedly during the long, dry season.

**Biomass Class O** only occurs in south-west Angola and parts of Botswana. It represents areas of bushy, arid vegetation adjacent to the deserts and salt pans. Productivity levels are very low and slightly seasonal.

**Biomass Class P** is indicative of deserts, salt pans and sand dunes, where vegetation productivity is severely limited.

**Biomass Class Q** is only found on the Mozambique–Swaziland border, and indicates relict patches of Lubumbo Hills Woodland. Biomass levels are high with a dip in productivity during the dry season.

**Biomass Class R** is only found in Lesotho where it represents degraded rangelands with invasive bush growth. It is similar to Class M but is less productive because of its high altitude and southerly location.

**Biomass Class S** is also found only in Lesotho and represents better quality rangelands. Grass productivity levels are moderate and woody vegetation is rare.

**Biomass Class T** is again restricted to Lesotho. It represents the low-productivity alpine and sub-alpine grass and heath vegetation.

**Biomass Class U** represents the Sublittoral Forests and Bushlands of southern Mozambique. These have moderate levels of productivity and a marked seasonality.

**Biomass Class V** is only found in southern Mozambique and indicates areas of Littoral Grassland with low woody biomass.

**Biomass Class W** is found in Lesotho and Swaziland. The class is characterized by areas of relict woodlands in the mountains. Productivity and biomass levels are high, but are restricted by low winter productivity.

**Biomass Class X** is only found in Swaziland and indicates the densest coniferous plantation stands. These are all in high-productivity commercial forests.

**Biomass Class Y** is found in Malawi and Swaziland. It represents extensive areas of non-coniferous plantations (e.g., *Eucalyptus* and *Gmelina*). These are high-productivity commercial forests.

**Biomass Class Z** is only found on the Swaziland Lowveld and represents the extensive irrigation schemes. Vegetation productivity levels are high but those of woody biomass are low.

The distribution of biomass classes and their growing stocks and productivity, are summarized in each country profile. Summaries for each administrative unit are also given in Chapter 13. It can be seen from Table 2.1 that slightly over a quarter of the sustainable woody reserves in the entire SADCC region are held in Angola. Angola is the largest country of the region, but it has vast areas of moist savannah woodland and transitional rain forest, both of which are highly productive.

Five countries can be identified with land areas of 60,000 to 80,000 km². Of these Mozambique, Tanzania and Zambia have similar levels of woody-biomass resource. Botswana, however, has significantly lower resources whilst falling in the same size class. Zimbabwe, with a smaller area than these four countries, has a higher growing stock than Botswana but it is lower than that of Mozambique, Tanzania or Zambia. All of the three smaller countries (Lesotho, Malawi and Swaziland) have lower woody-biomass resources than would be expected, despite their lower areas.

**Table 2.1**   SADCC: GROWING STOCKS AND MAI DISTRIBUTION BY COUNTRY

| Country | Area (km² × 1,000) (%) | Growing Stock (mill t × 100) | (%) | MAI (mill t) | (%) |
|---------|------------------------|------------------------------|-----|--------------|-----|
| Angola | 1246.70 | 47.13 | 26.9 | 143.20 | 25.8 |
| Botswana | 581.79 | 13.11 | 7.5 | 45.60 | 8.2 |
| Lesotho | 30.36 | 0.02 | <0.1 | 0.03 | <0.1 |
| Malawi | 117.41 | 3.68 | 2.1 | 13.71 | 2.5 |
| Mozambique | 799.38 | 35.15 | 20.1 | 114.00 | 20.5 |
| Swaziland | 17.37 | 0.24 | 0.1 | 1.33 | 0.2 |
| Tanzania | 840.10 | 33.61 | 19.2 | 107.04 | 19.3 |
| Zambia | 726.85 | 27.45 | 15.7 | 83.30 | 15.0 |
| Zimbabwe | 390.73 | 15.02 | 8.6 | 47.55 | 8.6 |
| TOTAL | | 175.41 | | 555.76 | |

mill t = million tonnes

25

| Woody-biomass resource supply potential | Any assessment of the woody-biomass supply potential of the biomass classes at the regional scale of this assessment depends on a consideration of the following: |

i)   growing stock;
ii)  annual and seasonal productivity;
iii) accessibility;
iv)  quality of wood available; *and*
v)   conflicting end uses.

Three factors have been considered when examining the accessibility of woody-biomass supplies in each of the classes. First, the ability to obtain small branches and twigs; secondly, physical access all year round; and thirdly, forest and woodland reservation.

For each country, each biomass class (and any major subdivisions) contains estimates of growing stock and annual productivity. Accessibility and wood quality are discussed wherever possible. This chapter provides only growing-stock and productivity estimates and comments on accessibility at regional level. Table 2.2 is a quantitative rating of the different biomass classes in terms of growing stock, annual productivity and accessibility.

**Table 2.2**  FUELWOOD SUPPLY POTENTIAL FOR BIOMASS CLASSES IN THE SADCC REGION

| Biomass Class | Growing Stock | Annual Productivity | Accessibility |
|---|---|---|---|
| A –Rain Forest/Miombo Woodland Transition | 71.22 | 2.25 | Good in places, but restricted by high crowns and reservations |
| B –Evergreen Miombo Woodland | 71.22 | 2.25 | Good in places, but restricted by high crowns and reservations |
| C –Wet Miombo Woodland and Montane Forest | 71.22 | 2.25 | Good in places, but restricted by high crowns and reservations |
| D –Seasonal Miombo Woodland | 19.85 | 0.49 | High crowns |

*Table 2.2 Continues*

*Table 2.2  Continued*

| | | | |
|---|---|---|---|
| E –Dry Miombo Woodland | 16.79 | 0.49 | Tobacco estates and timber operations |
| F –Cleared Miombo Woodland | 33.55 | 1.01 | Some areas have low wood: non-wood ratio |
| G –Coastal Forest | 88.78 | 4.98 | Flooding in mangroves |
| H –Dry and Degraded Miombo Woodland | 9.44 | 0.24 | Low productivity and low wood:non-wood ratio in places |
| I –Mopane Woodland | 36.97 | 1.09 | Low wood:non-wood ratio in places |
| J –(a) Commercial Agriculture | 0 | 0 | Restricted access and low woody biomass |
| –(b) Dry Savannah and Thickets | 11.83 | 0.42 | Low wood: non-wood ratio |
| K –Grassland and Bushland | 0–23.36 | 0–0.83 | Low wood: non-wood ratio |
| L –Bushland with Woodland and Scrubland | 14.22 | 0.60 | Low wood:non-wood ratio |
| M –Bushy Scrubland and Degraded Open Savannah | 7.11 | 0.30 | Low wood:non-wood ratio |
| N –Dry Scrubby Woodland and Grassland | 11.45 | 0.55 | Low wood:non-wood ratio |
| O –Arid Shrubland | 7.11 | 0.30 | Few restrictions |
| P –Arid Vegetation | 0 | 0 | Few restrictions |
| Q –Lubombo Hills Woodlands | 75.92 | 2.65 | Local access restrictions |

*Table 2.2 continues*

Table 2.2  Continued

| | | | |
|---|---|---|---|
| R – Grassland with Scrub Woodland | 0.50 | 0.01 | Low wood:non-wood ratio and reservations |
| S – Highveld and Riparian Grassland | 0 | 0 | Few restrictions |
| T – Alpine and Subalpine Grassland and Heathland | | | Few restrictions |
| U – Sublittoral Forest and Bushland | 9.44 | 0.24 | Few restrictions |
| V – Littoral Grassland | 0 | 0 | Few restrictions |
| W – Highveld Woodland | 28.08 | 0.30 | Mostly forest reserves |
| X – Dense Plantation Stands | 39.50 | 5.16 | All commercial forests, no access |
| Y – Plantations | 42.08 | 8.53 | All commercial forests, no access |
| Z – Irrigated Agriculture | 0 | 0 | All commercial farms, no access |

**Low woody-biomass supply areas**

The main areas with low woody-biomass supply are listed below. These are expanded upon in the country profiles (see Chapters 3–11) and only broad problem areas are indicated here. Italics are used to indicate areas of high population pressure and woody-biomass crisis. The effects of accessibility have not been taken into account here.

*Angola*

*Coastal plain, from the Namib Desert in the south-west to Cabinda Province; south-central Cunene Province;* the Zambezi Valley grasslands on the Zambian border.

*Botswana*

Most areas south of 20° south, particularly the *eastern half of the country.*

| | |
|---|---|
| *Lesotho* | *The entire country.* |
| *Malawi* | *Parts of the Central and Southern regions.* |
| *Mozambique* | *Southern Mozambique (parts of Gaza and Maputo Provinces)*; the Zambezi Valley (particularly in Tete Province); north-east Cabo Delgado Province; eastern Sofala Province. |
| *Swaziland* | *Southern and central Lowveld; central and northern Middleveld;* central and northern Highveld. |
| *Tanzania* | Central Tanzania (*excessive demand only in parts of this region, namely Shinyanga, Singida, Iringa and Mbeya Districts*); south-western Tanzania; *around Dar-es-Salaam; Usambara Mountains.* |
| *Zambia* | *Parts of north-eastern Zambia; the Line-of-Rail (from the Copperbelt through Lusaka and Kafue, to Livingstone).* |
| *Zimbabwe* | The Zambezi Valley and Escarpment in northern and north-eastern Zimbabwe; *the Limpopo and Sabi Valleys; communal areas on the Highveld to the south of Harare and in Masvingo and Matabeleland South Provinces.* |

# 3. Angola

GENERAL
DESCRIPTION AND
BIOMASS CLASSES

Angola can be divided into 12 biomass classes (see map on p.33) defined by vegetation phenology and productivity. These classes are based on the interpretation of NOAA 7 AVHRR GAC data and previous botanical and forestry research (Coelho, 1967; Huntley, 1974; Menzes; 1971, Monteiro, 1970; Monteiro & Sandinha, 1971).

The standing biomass of the different vegetation types is slightly less important in this classification than the phenology and productivity. The floristic composition is relatively unimportant. Consequently, some biomass classes incorporate a variety of vegetation types. They are related in terms of phenology, productivity and biomass, which have been identified previously as important by botanists (e.g., Barbosa, 1970; White, 1983), as distinct from floristic units. These biomass classes are more appropriate for the appraisal of fuelwood resources than the previously defined floristic-mapping units.

Standing biomass reserves are very high over much of the country. Four of the 12 biomass classes identified and described below are dominated by forest and woodland and have high fuelwood and timber potential. They also account for 61.6% of the country. This is substantially higher than the 40% closed forest and savannah-woodland cover calculated by Lanly (1981); even if his deforestation rate of 40,000 ha yr$^{-1}$ is taken into account. One reason for this discrepancy is that this study has included the wooded savannahs and cleared areas in the north of the country in the four forest and woodland biomass classes. Nevertheless, it is still difficult to arrive at Lanly's estimate and it appears that Angola has considerably more forest and woodland than has been previously suggested. Added to this, some of the dry-savannah biomass classes have high standing woody biomass as well. Three classes – Coastal and Desert Vegetation and the two grassland classes – are characterized by very low fuelwood reserves (5.61% of the country by area). Biomass classes dominated by savannah vegetation have intermediate fuelwood reserves and they account for 29.65% of the country by area. The wide variation in vegetation types, and therefore fuelwood potential, is due to the fact that much of northern Angola is transitional with the rain forest zone, and the south-west is in the arid zone. Despite this extreme range of vegetation types, much of the country is covered by miombo woodland, varying from Dense, High Miombo Woodland to Dry, Open, Deciduous Miombo Savannah. The strongest controls on miombo woodland distribution in Angola appear to be vegetation disturbance, rainfall amount and seasonality, depth of soil, and altitude.

Despite the very healthy national biomass picture, there are a number of areas that have obvious problems with fuelwood supply. The districts most affected, in terms of the areal extent of moderately and highly productive fuelwood sources, are:

- Bengo
- Namibe
- Luanda

These areas have been identified previously. Trees in Bengo and Luanda Districts have been cleared for agriculture, grazing and fuelwood at an increasing rate over the past 15–20 years and the situation in these districts is now critical. The Namibe District suffers from a natural shortage of productive biomass reserves and the problems of fuelwood supply to the main towns of Bibala, Namibe, Tombwa and Virei are acute. Estimates have shown that the cutting of vegetation exceeds annual growth in Namibe District. There is evidence that the Dry Deciduous Savannah in Huila District, along the Lubango–Huambo Road and the Lubango–Namibe Railway, is being exploited to meet the fuelwood demands in Namibe. In addition, more localized shortfalls in production can be recognized from the biomass-class maps and productivity data for the following districts:

i) Benguela – along the coast and inland as far as Boccoco and Cantangue;
ii) Cabinda – along the coastal plain;
iii) Cunene – around N'Giva;
iv) Kwanza Norte – in the western half of the district;

v)   Kwanza Sul – along the coast and inland as far as Gabela
      and around Quibala;
vi)  Moxico – along the Luena Valley; *and*
vii) Zaire – along the coastal plain.

The vegetation zones into which Angola has been classified are:

Transitional Rain Forest/Miombo Woodland
Dense, High Miombo Woodland
Dense, Medium-Height Miombo Woodland
Seasonal Miombo Woodland and Wooded Savannah
Dry Deciduous Savannah
Dry Coastal Savannah and Arid Coastal Thicket
Dry Inland Savannah
Degraded Rain Forest and Miombo Woodland
Bushy Arid Shrubland
*Chanas da Borracha* Grassland and Degraded Dry Deciduous
   Savannah
Montane Grassland
Coastal and Desert Vegetation

Throughout the report the districts of Bengo and Luanda have been dealt with as one, under the name of Bengo.

## DESCRIPTIONS OF INDIVIDUAL BIOMASS CLASSES

### Transitional Rain Forest/Miombo Woodland

The vegetation related to the transition between the rain forest to the north and the miombo-woodland belt to the south, covers extensive areas of north-central and north-western Angola. The largest of these are in Malanje, Lunda Norte, Lunda Sul and Uige Districts; but smaller areas are found in Cabinda, Kwanza Sul, Moxico and Zaire Districts. This vegetation class is related to both the forests of the Zaire Basin and the miombo woodlands of the Angolan plateaux and has many ecological, floristic and phenological similarities with them both. In actual fact, significant areas of miombo woodland in Bie and Huambo Districts have been classified with the related transitional vegetation to the north, on the basis of phenology and productivity. The total area covered by this biomass class is 159,680 km$^2$ (that is 12.8% of the country).

This vegetation exhibits high levels of photosynthetic activity and a marked seasonality. NDVI values are high from November until April, NDVI being about 180 to 185. From May until July NDVI declines rapidly, reaching 150 in July and August. This period

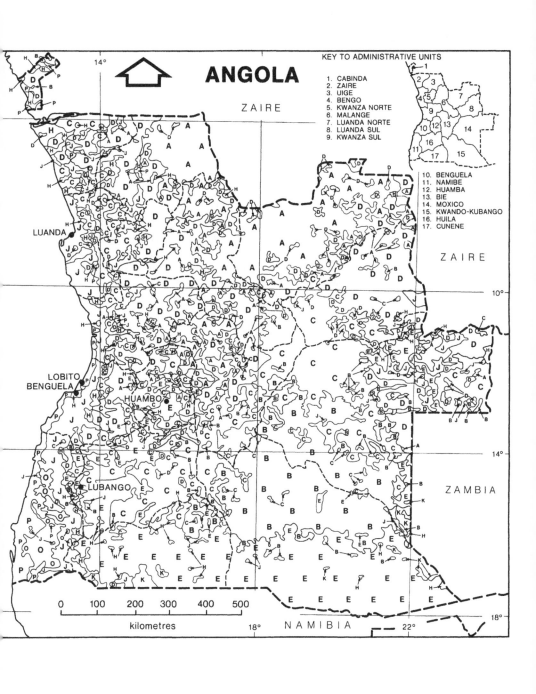

ANGOLA

ZAIRE

KEY TO ADMINISTRATIVE UNITS

1. CABINDA
2. ZAIRE
3. UIGE
4. BENGO
5. KWANZA NORTE
6. MALANGE
7. LUANDA NORTE
8. LUANDA SUL
9. KWANZA SUL

10. BENGUELA
11. NAMIBE
12. HUAMBA
13. BIE
14. MOXICO
15. KWANDO-KUBANGO
16. HUILA
17. CUNENE

LUANDA

ZAIRE

LOBITO
BENGUELA

HUAMBO

LUBANGO

ZAMBIA

0   100   200   300   400   500

kilometres

NAMIBIA

of low rates of photosynthetic activity is followed by a rapid rise in November.

The forest on the southern fringes of the Zaire Basin differs from the rain forest found in the central basin. The coastal area is much drier and its marked seasonality makes the forest semi-evergreen. It has been described by White (1983) as "drier peripheral semi-evergreen rain forest".

Much of this peripheral rain forest has been extensively cleared in the past and, outside forest reserves, it is debatable whether any primary forest remains. Nevertheless, undisturbed mature secondary rain forest has similarities with primary forest. Both categories are most highly developed in the undisturbed parts of the Dembos Cloud Forest (dealt with in the Dense, Medium-Height Miombo Woodland Forest biomass class); in inaccessible parts of Malanje Plateau; and in the north-south trending river valleys which dissect the northern plateaux. It includes a variety of vegetation types related to the rain forest and is influenced strongly by anthropogenic activity, particularly bush-fallow agriculture. These vegetation types range from mature secondary and primary forest, through various stages of secondary forest regrowth intermixed with miombo woodland (the dominant vegetation type), to wooded grassland in young fallow and badly degraded areas.

Areas of mature forest are floristically rich and structurally well defined, but they are confined to the river valleys and inaccessible parts of the Malanje Plateau. They have a closed, upper canopy of 30–50 m which blocks the passage of solar radiation to the ground. As a consequence, the under-storey tree and shrub layers are poorly developed. These layers consist of:

    i)   a less dense shrub layer that is floristically related to the upper canopy and reaches 8 m in height;

    ii)  a poor herb layer; *and*

    iii) a few lianas and epiphytes.

The way in which secondary forest develops after disturbance depends on the level of that disturbance, the rainfall regime, and soil properties. In the less disturbed wetter areas the succession can be divided into three stages, all of which have different woody-biomass potential. The pioneer secondary forest is dominated by bushes, small woody shrubs and small trees varying in height from 4 to 12 m. After three to four years the bushy stage is replaced by a young, secondary forest which lasts for about a decade. This consists primarily of *Musanga cecropiodes* and other trees which have emerged from the species that dominated the early bushy stage. The final phase, prior to maturity, is characterized by a large number of rapidly growing trees reaching heights of 35 m.

A more common succession is found on the drier, interior plateaux of northern Angola. These plateaux are heavily dissected by north-south trending rivers. These river valleys contain a rapidly regenerating riparian forest that is 35–40 m tall. On the plateau, however, the forest is less productive and more sensitive to disturbance. The most extensive forest development is of dry evergreen forest (known as *mabwati* in Zaire). This consists of a dense evergreen canopy of about 25 m tall, consisting mainly of *Brachystegia spiciformis, B. waangermeeama, Isoberlinia giorgii, Lannea antiscorbutica, Marquesia acuminata, M. macroura,* and *Parinari curatellifolia.* Beneath it, there is a shrub layer (dominated by *Hymenocardia* spp., *Maprounea* spp. and *Psorospermum* spp.) and a tall well-developed layer of grass. The dry evergreen forest has been subject to much burning and clearance over a long period of time. Where this has not occurred miombo woodland has developed. However, under more severe disturbance the vegetation reverts to wooded grassland.

Miombo woodlands are found throughout this biomass class. In north-west Lunda Norte Province, 35–40 m semi-evergreen *Marquesia acuminata–Pteleopsis diptera* woodland is found, but *Marquesia macroura-Brachystegia taxifolia* woodland is more common. This is a successional miombo woodland with a 30 m dense, evergreen canopy of the two forementioned species, with *Brachystegia* spp., *Daniellia alsteeniana, Pterocarpus angolensis,* and *Uapaca* spp. There is also a small tree and shrub layer dominated by *Bridelia* spp., *Erythrophloeum* spp., *Faurea saligna, Parinari curatellifolia,* and *Uapaca* spp. It is likely that the miombo woodlands in Bie and Huila Districts, classified as belonging to this biomass class, are similar to the woodlands in Lunda Norte District.

In areas of more severe deforestation, the woodland only recovers to form a wooded grassland known as *mikwati* in Zaire. It consists of a tall, well-developed grassland with scattered individual, fire-resistant trees such as *Burkea africana, Combretum celastroides* subsp. *laxiflorum, Dialium engleranum, Diplorhynchus condylocarpon, Erythrophloeum africanum, Hymenocardia acida, Protea petiolaris* and *Pterocarpus angolensis.*

Obviously, the fuelwood potential of this biomass class is variable. If the most extensive forest and woodland types are compared, there is probably little difference between them and miombo woodland, even following disturbance. However the fuelwood potential of the wooded grassland is quite low, both in terms of standing stock and annual productivity.

**Dense, High Miombo Woodland**

Dense, High Miombo Woodland is contained within an extensive area of about 111,281 km². It is mainly found in Bie, Huila, Kuando-Kubango, Cunene, and Moxico Districts. The biomass class accounts for about 8.9% of the land cover, and 65.1% is found in Kuando-Kubango District. It mainly occurs on slightly undulating sandy plains associated with the headwaters of the Cuito, Kuando, Kubango, Luanginga, Luena, and Lungue-Bungo Rivers to the south-east of the Bie Plateau.

The phenology of Dense, High Miombo Woodland in Angola shows very little seasonal variation in productivity. In fact, this area of woodland is one of the most seasonally resilient in the entire SADCC region. Productivity is high until mid-May (with NDVI values ranging between 175 and 185); then there is a slight drop with the lowest values (NDVI of about 160) occurring in September and October; after this, productivity increases again. It can be concluded that this type of miombo woodland suffers very little moisture stress and is essentially evergreen.

Dense, High Miombo Woodland has a well-defined structure and is floristically rich. The canopy, which is often as high as 25 m, is dense, closed, and dominated by *Brachystegia bakerana, Crytosepalum exfoliatum, Dialium engleranum, Guibouria coleosperma* subsp. *pseudotaxa,* and *Julbernardia paniculata.* There are few other canopy trees, but stunted trees grow to between 5 and 10 m. This lower layer is dominated by *Baphia massaiensis, Copaifera baumiana, Diospyros* spp., *Landolphica* spp., *Paropsia brazzeana, Parinari* spp., *Trichila quadrivalis* and *Xylopia odoratissima.* In addition to these two tree layers, there is a very poorly developed grass and herbaceous ground layer.

Detailed data on the composition of Dense, High Miombo Woodland for Kuando-Kubango Districts are available in the forestry feasibility study carried out by Coelho (1967). This area is mainly of Dense, High Miombo Woodland and was divided into 25 forest zones. A variety of data were collected and the salient information was tabulated (see Table 3.1).

Here, Dense, High Miombo Woodland differs slightly from that to the north as trees related to the Dry Savannah Woodland (which interdigitates with this biomass class) are to be found, most commonly, *Burkea africana*.

The Dense, High Miombo Woodland is dominated here by *Brachystegia bakerana Cryptosepalum pseudotaxius, Erythrophleum africanum, Guibourtia coleosperma* and *Isoberlinia* spp. (especially *I. baumii*). On the shallower soils of the Bie Plateau, and close to the savannah grasslands to the south, the nature of the canopy changes to a more open form.

**Table 3.1** SUMMARY DATA OF FOREST COMPOSITION IN THE DENSE, HIGH MIOMBO WOODLANDS OF KUANDO-KUBANGO DISTRICT

| Trees | | No. of occurrences as a dominant in biomass classes | Ranges | |
|---|---|---|---|---|
| Species name | Local name | (max = 25) | Trees/ha (dbh >20 cm) | Frequency of occurrence (%) |
| *Brachystegia longifolia* | Mussamba | 4 | 0–10 | 0–13 |
| *B. bakerana* | Mutete | 5 | n/d | 13–32 |
| *B.* spp. (unknown) | Mumango | 2 | 4.9–14.6 | 6–13 |
| Other *Brachystegia* | – | 5 | 0.5–16 | 2–19 |
| *Isoberlinia baumii* | Mumue | 15 | 3.4–24 | 1–68 |
| *I.* spp. (unknown) | Mussonda | 1 | 3.2 | 5 |
| Other *Isoberlinia* | – | 4 | 8–49 | 3–41 |
| *Guibourtia coleosperma* | Mussibi | 15 | 2–29.4 | 5–46 |
| *Crypotsepalum pseudotaxus* | Mucuve | 4 | n/d | 7–35 |
| *Pterocarpus angolensis* | Girassonde | 4 | 0.3–2.2 | 3–10 |
| *Burkea africana* | Mucesse | 7 | 1–3 | 6–24 |
| *Combretum* spp. | Mupoloti | 2 | 2 | 4–5 |
| *Diplorhynchus angolensis* | Muuri | 1 | n/d | 6 |
| *Erythrophleum africanum* | Mucosso | 15 | 2–29.4 | 0.5–4.46 |
| *Ochna* spp. | Mufuco | 2 | n/d | 5–7 |
| *Monotes africanus* | Mucenene | 4 | n/d | 8–17 |
| *Ricinodendron sautaneii* | Mangongo | 1 | n/d | 13 |

Adapted from Coelho (1967).
n/d = no data

In this class, standing biomass is very high and annual productivity is generally high throughout. Consequently, there are considerable fuelwood reserves in the districts covered by this biomass class.

## Dense, Medium-Height Miombo Woodland

Dense, Medium-Height Miombo Woodland differs from Dense, High Miombo Woodland in that it has slight seasonal differences and a lower overall level of productivity. This is partly due to the fact that it is mainly developed on the drier, sandier and stonier soils to the north and west of the Dense, High Miombo Woodland on the Bie Plateau. It is also found in restricted areas of eastern Moxico District where it is closely related to the miombo woodlands of Zaire and Zambia. It covers very large portions of Bie (32.6%), Huila (56.3%), Lunda Sul (61.2%) and Moxico (23.8%) Districts. There are also extensive areas in Cunene, Bengo, Huambo, Kuando-Kubango, Malanje, and Zaire Districts.

Semi-deciduous Dembos Cloud Forest is also included in this biomass class. It is found on the main escarpment in Kwanza Norte District. Together, these two vegetation types account for 221,164 km² and make up the second largest biomass class covering 17.7% of Angola.

It can be phenologically distinguished from Dense, High Miombo Woodland by a marked period of moisture stress which reaches a maximum in August, and the semi-deciduous nature of the woodland. Generally, productivity levels are high and wet-season NDVI values range from 175 to 185.

Like its evergreen equivalent, Dense Medium-Height Miombo Woodland has a well-defined structure and is floristically rich. There is a dense, relatively closed canopy which often exceeds 15 m and can reach 30 m. It is dominated by *Brachystegia boehmii., B. gosswieleri, B. spiciformis, B. wangermeeana, Combretum* spp., *Cussonia angolensis, Isoberlinia angolensis* and *Julbernardia paniculata*. Other canopy trees are less common but there is a well-developed stunted tree layer of between 5 and 10 m. There is also a well-developed herbaceous undergrowth. Within the woodland, 4–10 m high stands of *Uapaca* forest are found. These are indicative of a successional vegetation following agricultural clearance.

On the higher ground in the Bie Plateau, the woodland canopy is lower (6–25 m) and more open (cover varies from 30–90%). Here, it is dominated by *Brachystegia bakerana, B. longifolia, B. spiciformis, Copaifera baumiana* and *Guibourtia coleosperma*. There is a low (0.5–2 m) woody layer with 50–60% ground cover and a dense, tall (2–3 m) grass layer. The fuelwood potential of the miombo woodland on the Bie Plateau is generally high, both in terms of growing stock and annual productivity. The exceptions are the woodlands developed on thin soils, in areas of high ground, where the growing stock is low. Nevertheless, the annual productivity in these areas is still relatively high.

The woodlands falling into this biomass class in eastern Moxico District are ecologically and floristically related to the woodlands of southern Zaire and western Zambia. They are highly variable and closely related to the depth and quality of the soil. On well-drained soils, various types of miombo and *Cryptosepalum* woodland have developed.

On the thinnest soils these are dominated by *Brachystegia microphylla* and *B. utilis*. Closed *B. spiciformis* woodlands are also found and, on the thicker well drained soils, there is either open *Marquesia calonerus-Uapaca pilosa* woodland or, more commonly, dense *Cryptosepalum maraviense* woodland. All of these woodland types have high canopies (15–25 m) but vary considerably in canopy openness and the amount of under-storey vegetation. On the poorly drained soils, the woodland is much lower in height and is dominated by *B. boehmii*, *Marquesia katangensis*, and *Uapaca* spp. In many places this low woodland develops into a grassland with very low fuelwood potential. These eastern woodlands have highly variable fuelwood potential that is dependent on local environmental conditions.

The Dembos Cloud Forest, found on the north-eastern escarpment of the central plateau between 6°30′N and 9°30′N, is also included in this biomass class. Cloud forest vegetation exhibits less moisture stress than that on the adjacent lowlands and plateau. This is due to the frequent mists and increased precipitation (annual rainfall varies from 1,100 to 1,500 mm) from the moisture-laden onshore winds. The forest canopy reaches between 30 and 50 m and is floristically rich. The main canopy trees are *Alibizia* spp., *Celtis* spp., *Ficus* spp. and *Morfus* spp.

The Dembos Cloud Forest is used extensively for coffee cultivation. The under-storey trees are destroyed and coffee bushes planted in the shade of the tall canopy trees. This reduces the total biomass and so, although this area is floristically related to the rain forest, it is classified with the denser miombo woodlands.

A more important consequence of the under-storey destruction is the reduction in accessible fuelwood trees, and therefore in the

fuelwood growing stock. The problem of diminishing poten
fuelwood biomass is exacerbated by the planting of coffee and
restricted access this cultivation imposes on fuelwood collector

## Seasonal Miombo Woodland and Wooded Savannah

This class includes a number of closely related types of vegetati
ranging from miombo woodlands and bushy thickets to w
wooded savannahs on the plateaux. All are characterized b
strong seasonality and a deciduous nature. These vegetation ty
merge into each other and the surrounding woodlands; they
found extensively in Angola, bordering the rain forest to the no
and the denser miombo woodlands to the south.

The main areas in which this class is found are Kwanza S
Lunda Norte, Lunda Sul, Malanje and Moxico Districts. Sligh
smaller areas are found in Bie, Kwanza Norte, Uige and Za
Districts. In addition, there are small areas in other districts, m
notably Huambo and Huila. The total area is 306,946 km² wh
accounts for nearly a quarter of the country (24.6%) making it
most areally extensive class. On a district basis it accounts for o
half of Kwanza Norte (51.9%), Kwanza Sul (59.6%), Lunda
(60.7%), Malanje (57.1%), and Zaire (50.2%) Districts.

Seasonal Miombo Woodland is intermediate between se
deciduous Dense, Medium-Height Miombo Woodland and
drier, deciduous woodlands. This relationship is clearly illustra
by its phenology. In the wet season the rates of photosynthe
activity are similar to the Dense, Medium-Height Miombo Wo
land (NDVI varies between 180 and 185). Rates then decline rap
until the dry season, when rates of photosynthesis are compara
to Dry, Deciduous Savannah (NDVI varies between 145 and 15

The western part of this biomass class (west of Cangumbe),
Bie, Kwanza Norte, Kwanza Sul, Malanje, Uige and Zaire Distri
is dominated by wooded savannah. Depending on the lo
environment, there are also patches of forest, bushy thickets a
grassland. It is drier and more deciduous than the adjacent for
and woodland, and differs floristically from the miombo woodlan
The dominant species are *Acacia sieberana, Annona* spp., *Cochlosperm
angolense, Diplorhynchus condylocarpon, Piliostigma thonningii* and *Termin
sericea*. The shrub layer can develop into a thicket-like savannah
there is usually only a well-developed tall grass cover.

The eastern part of the biomass class is characterized by an op
canopy miombo woodland which is floristically poorer than Den

Medium-Height Miombo Woodland. Most notably, *Brachystegia floribunda* is far less common and in some areas is absent altogether. The other species of *Brachystegia* found in the wetter miombo woodlands of Angola are less important; in most areas, particularly on thin or free-draining soils, *Brachystegia gosswielei, B. spiciformis* and *Julbernardia paniculata* dominate. The greater penetration of light through the canopy has resulted in a larger stunted, shrubby under-storey of trees; and a better developed grass and herbaceous ground layer. Nevertheless, in wetter areas, the canopy trees are very similar to the two wetter miombo woodland types.

In northern Bie District, Seasonal Miombo Woodland is very poorly developed. The canopy varies between 6 and 30 m in height and the tree canopy cover varies from 15 to 70%. The canopy is dominated by *Brachystegia tamarindoides, Marquesia loadensis, Syzygium guineensee* and *Uapaca* spp. The openness of the canopy leads to a low tree and shrub layer that reaches 5 m, and an open grassy undergrowth.

The distinct seasonal dip in productivity, combined with the open nature of much of the woodland and savannah in this biomass class, means that the potential for fuelwood production is lower here than in the other miombo woodlands on the plateau. There is also a markedly lower fuelwood potential, particularly in terms of growing stock, in the eastern part of the biomass class (that is in Bie, Kwanza Norte, Kwanza Sul, Malanje, Uige, and Zaire Districts) compared to the western part (Lunda Norte and Lunda Sul Districts).

## Dry Deciduous Savannah

Dry Deciduous Savannah is found in the southern half of Angola and is particularly common along the Namibian border. It also occurs extensively on the Kalahari Sands of the upper Zambezi Basin (extending into Zambia) and on the highest parts of the Bie Plateau.

This biomass class includes the driest variants of the Angolan miombo woodlands as well as the *Baikiaea* and *Mopane* woodlands. These usually take the form of a wooded savannah. Areally extensive tracts of Dry Deciduous Savannah occur in Cunene, Kuando-Kubango and Moxico Districts, but there are also locally important areas in Huambo, Huila, and Namibe Districts. It is the most extensive biomass class in Cunene, Kuando-Kubungo, and Moxico Districts, covering 63.8%, 43.7% and 33.9% of these districts respectively. It covers a total area of 229,657 km² or 18.4% of the country.

Dry Deciduous Savannah has a far lower annual productivity than the woodlands in Angola and is strongly seasonal. Wet season productivity peaks between February and April, when NDVI ranges from 170 to 175. This is followed by a slow decline and the period of lowest productivity is September when NDVI is approximately 150.

The seasonal drought in the south and east is due to a low annual rainfall and the long dry season. This seasonality is accentuated by the low water-holding capacities of the soils developed on the Kalahari Sands and the thin stony soils of the Bie Plateau. The inability of these soils to hold water throughout the dry season means that the trees are subject to a severe seasonal drought.

Dry Deciduous Savannah is both structurally and floristically distinct from the plateau woodlands. It is also more open. The canopy trees are generally low (about 10 m) and are interspersed with a grass layer of up to 2 m in height. On the Bie Plateau, the dominant trees are *Brachystegia boehmii, B. gosswielei, B. spiciformis, Julbernardia globiflora* and *J. paniculata*. On the high parts of the plateau, between 1,900 – 2,200 masl (metres above sea level), the canopy is much lower (not more than 5 m) and dominated by *B. floribunda, B. spiciformis* and *J. paniculata*. However, a variety of other shrubby and stunted trees has invaded the high ground to form a bushy thicket. Above 2,200 masl, the woody species die out and the open thicket is replaced by montane grassland. The fuelwood potential of these woodlands is far lower than that of others in Angola, with only moderate levels of productivity and growing stock. On the highest parts of the plateau the fuelwood potential is extremely low, but fortunately these areas are quite localized.

In Kuando-Kubango District, in south-east Angola on the Kalahari Sands, where the rainfall varies between 500 and 1,000 mm, *Baikiaea* woodland and *Burkea africana* savannah is found. These vegetation types were also included in the forest feasibility study of Kuando-Kobango (Coelheo, 1967). Seven forest zones were identified in the area covered by this biomass class. Twelve species dominated the forest plots, some of which only occurred in one or two zones and were obviously related to the miombo woodlands to the north (see Table 3.2).

The main dominants are characteristic of this biomass class; they are *Burkea africana, Baikiaea plurijuga, Pterocarpus angolensis, Combretum dinteri* and *Guibourtia coleosperma*. Tree densities of the dominant species range from 1.5 to 47% for all trees, with 1.4 to 48% for trees of greater than 20 cm dbh. *B. plurijuga* forms a canopy which can reach 20 m, but is usually lower in the drier areas. Beneath the canopy is a rich, dense shrubby under-storey. The *Baikiaea* woodlands grade into the Dense, High Miombo Woodlands to the north, and the Mopane Woodlands to the west. *B. plurijuga* is an important timber tree and consequently there is a potential conflict in terms of fuelwood utilization. However, the under-storey has a high growing stock of more accessible tree and shrub species. Consequently, the fuelwood potential of much of the *Baikiaea* woodland is quite good, although in the drier areas it can be restricted by low annual productivity.

Mopane Woodland occurs on the alluvial soils of the Cunene Valley in south-west Angola, where precipitation varies between 400 and 600 mm. It usually forms an open woodland with a tree canopy varying from about 15 m in the west, to 7 m in the east. This height difference is a response to the lower level of rainfall in the east. The canopy is dominated by *Acacia kirkii, Colophospermum mopane* and *Sterculia* spp. There is a sparse under-storey of grasses, herbs and small trees, but the ground cover rarely exceeds 50%. In disturbed areas the trees are soon reduced to a low, bushy habit or grassland. The fuelwood potential of Mopane Woodland is the lowest of all the Angolan woodlands, both in terms of growing stock and annual productivity.

**Dry Coastal Savannah**

The Dry Coastal Savannah biomass class extends along much of the Angolan Coastal Plain and the foot of the Escarpment in a strip which varies in width from 20 to 150 km. Annual rainfall in this area is no more than 600 mm.

This biomass class is floristically heterogeneous and includes low, degraded forms of miombo and mopane woodlands (usually in the form of a low-canopy woodland), grass-dominated savannah, thicket, and scrub formations. All of the vegetation types are adapted to a long seasonal moisture stress and, consequently, grasses are common. The Arid Coastal Thicket of west-central Angola is also found in this biomass class. It is areally important in Bengo, Benguela, Kwanza Sul and Zaire Districts. In total it accounts for 48,484 km², only 3.9% of the total land area.

Despite the variations in vegetation, all of the dry savannah communities show strong similarities in their phenology. Season-ality is not very marked and the wet-season productivity level is lower than in surrounding woodlands, the maximum NDVI being about 170. Productivity is lowest in September and October when the NDVI varies from 150 to 155.

In west-central Angola, between the Benguela-Namibe border and N'Zeto (in Zaire District) this vegetation has been called "north Zambezian undifferentiated woodland and wooded grassland" (White, 1983). It is floristically richer than the miombo and mopane woodlands and the dominant tree species of those two types of woodland are absent. Instead it is dominated by *Adansonia digitata, Dichrostachys* spp., *Euphorbia conspicua, Setaria welwitschii* and *Sterculia setigera*. Within the wooded grasslands there are variations in the vegetation structure which affect the fuelwood potential. In some areas, grasslands are dominant and the woody component is restricted to small islands of *Acacia* spp., *Dichrostachys cinerea, Euphorbia conspicua* or *Terminalia prunioides*. Other areas of swampy grassland and forest occur along the rivers cutting through the coastal plain. In stony areas, the woody component is restricted to a shrubby thicket dominated by *Croton* spp., *Grewia* spp. and *Strychnos henningsii*.

The fuelwood potential of all the vegetation types within the Dry Coastal Savannah and Arid Coastal Thicket biomass class is low. The most productive areas are along the rivers but there are problems in gaining access to them. Vegetation types on the higher ground have varying growing stocks, but in all cases the annual level of productivity is low.

**Dry Inland Savannah**   Areas of Dry Inland Savannah are found in Cunene and Kwanza Norte Districts. They have strong similarities to the Dry Coastal Savannahs, in terms of biomass parameters, but exhibit floristic differences. In total they account for 26,263 km², that is about 2.1% of the total land area. Productivity levels are moderately low, with NDVI values varying from 170 in the wet season to about 150 in September and October.

There is disturbed woodland in the south, between the Namibian border and the River Coporolo. This is a transitional zone between the Bushy Arid Shrubland that lies to the west and open Mopane Woodland to the east. This area is far less disturbed than the areas to the north: the increasing aridity on the western edge of the Mopane Woodland has resulted in open wooded grassland

**Table 3.2** SUMMARY DATA OF FOREST COMPOSITION IN THE DRY DECIDUOUS SAVANNAH WOODLANDS OF KUANDO-KUBANGO DISTRICT

| Trees | | No. of | Ranges | |
| | | occurrences as a dominant in biomass classes | Trees/ha | Frequency of occurrence |
| Species name | Local name | (max = 7) | (dbh >20 cm) | (%) |
| --- | --- | --- | --- | --- |
| Burkea africana | Mucesse | 6 | 1.7–3.5 | 7–47 |
| Baikiaea plurijuga | Muiumba | 4 | 6–48 | 1.5–26.4 |
| Pterocarpus angolensis | Girassonde | 4 | 1.4–11.5 | 4–23 |
| Combretum dinteri | Mupopo | 3 | n/d | 3–20 |
| Guibourtia coleosperma | Mussibi | 3 | 2.4–4 | 3–12 |
| Erythrophleum africanum | Mucosso | 2 | 1.1–1.7 | 3–8 |
| Ochna spp. | Mufuco | 2 | n/d | 7 |
| Brachystegia bakeriana | Mutete | 1 | n/d | 13 |
| Combretum spp. | Mupoloti | 1 | n/d | 5 |
| Isoberlinia baumii | Mumue | 1 | 8.3 | 21 |
| Ricinodendron rautaneii | Mangongo | 1 | n/d | 23 |
| Terminalea sericea | Mangongo | 1 | 1.4 | 9 |

Adapted from Coelho (1967).
n/d = no data

vegetation, a canopy that is far more open than it is further east, and a well-developed layer of grass and herbs. Floristically, this area is similar to Mopane Woodland and the canopy is dominated by *Acacia kirkii*, *Colophospermum mopane*, and *Sterculia* spp.

The dry savannah woodlands of Kwanza Norte are intermediate between the drier and degraded forms of miombo woodland on the plateaus, and the dry coastal savannah vegetation types to the west. The fuelwood potential of this biomass class is restricted by the low annual productivity, although at a local level growing stocks may be high. The degraded nature of some of the areas means that further disturbance may inflict an irreversible decline on woody-biomass resources.

**Degraded Rain Forest and Miombo Woodland**

In areas where there has been extensive disturbance, the rain forest and miombo woodland are unable to regenerate properly and a secondary grassland or wooded grassland dominates the area. This biomass class represents such degraded areas which are scattered throughout the country. The main areas of degraded rain forest are found along the coastal plain, betwen N'Zeto and the Zaire River; among the inland areas of Kwanza Norte, Malanje, Uige and Zaire Districts; and on the coastal plain of Cabinda. Degraded Miombo Woodland is mainly found in Benguela, Cunene, Huila, Kwanza Sul and Uige Districts. In total, these areas cover about 33,220 km² (18,002 km² of Degraded Miombo Woodland and 15,218 km² of Degraded Rain Forest). This represents about 2.7% of the land area.

The vegetation is characterized by a relatively constant level of productivity between January and May with NDVI reaching about 175. After May, there is rapid decline in productivity and the NDVI reaches a dry-season low of 135 between July and October, after which it rises again.

In the degraded rain forest areas, the grass layer reaches heights of about 2 m. Individual fire-resistant trees and bushy thickets are found within the grasslands. The most common tree species are *Acacia welwitschii*, *Adansonia digitata*, *Euphorbia conspicua* and *Sterculia africana*; cashew and mango trees are also commonly found. In some areas the trees become stunted and a shrubby thicket develops; particularly common in these situations is *Strychnos henningsii*.

Degraded Miombo Woodland has a similar vegetation structure, phenology, growing stock and productivity to degraded rain forest. Apart from the introduced fruit trees such as cashew and mango, however, it is floristically different. The proportion of dominant miombo woodland trees (*Brachystegia* spp., *Isoberlinia* spp., *Julbernardia* spp. and *Mucronata* spp.) is much lower, and in some areas they are absent altogether. In these situations they are replaced by such trees as *Parinari curatellifolia*, *Monotes* spp. and *Uapaca* spp.

The fuelwood potential of the Degraded Rain Forest and Miombo Woodland is low because only a small proportion of the woody biomass is accessible and its annual productivity is not high. Gaining access to suitable wood is a problem in this biomass class as the branches of many of the trees are high in the crowns and a number of the important trees (e.g. cashew and mango) have uses other than fuelwood.

**Bushy Arid Shrubland**

Bushy Arid Shrubland is found extensively on the fringes of the Karro and Namib Deserts. It extends northward from the Namib Desert into southern Angola and is locally important in south-eastern Namibe District. It only covers 15,748 km² and in total accounts for 1.3% of Angola. It does, however, cover 22.7% of the Namibe District.

Bushy Arid Shrubland is dominated by small thorny bushy trees, many of which are succulent. They are scattered in a low grass and herb cover, giving the impression of a poor savannah. Productivity levels are very low throughout the year but there is a slight rise between March and August. NDVI values never exceed 140. For the rest of the year, productivity is low with NDVI varying between 125 and 130.

The main woody trees and shrubs are *Acacia* spp., *Boscia albitruncata*, *Commiphora* spp., *Grewia bicolor*, *Hoodia currori*, *Rhigozum* spp., *Salvadora persica* and *Sterculia* spp. The woody component rarely exceeds 2 m in height. As the Bushy Arid Scrubland appproaches the escarpment in the east it becomes denser and invasive species from the adjacent Mopane Woodland are found. The fuelwood potential of this biomass class is very low. It is restricted by the low growing stock, a low wood : grass ratio and a low annual level of productivity. Consequently, there is little potential for fuelwood exploitation other than at a local level.

The fuelwood-supply problems of Namibe District need to be put in perspective at this stage. This shrubland of low-growing stock and little productivity contains most of the woody vegetation in Namibe District but is only covers a quarter of the area. Most of the remaining land is desert.

**Chanas da Borracha Grassland and Degraded Dry Deciduous Savannah**

*Chanas da Borracha* Grassland is restricted to the upper Zambezi Valley on the Kalahari Sands in Kuando-Kubango, Moxico and Lunda Sul Districts. Degraded Dry Deciduous Savannah has similar biomass properties to the grasslands on the Zambian border. Areas of this type are found on the Namibian border near N'Giva, in Cunene District and in Kuanda-Kabango District. These two types of vegetation account for 72,238 km² or 5.8% of the country.

Levels of vegetation productivity show slight variations during the year but for most of the time are low with NDVI values ranging from 145 to 150. There is a slight increase in productivity between April and June, when NDVI values reach 160.

The grassland in the upper Zambezi Basin is known as *chanas da borracha* and is edaphically controlled by annual burning and the seasonal flooding of the soils developed on the Kalahari Sands. Its distribution closely follows the valleys of the rivers draining into the Zambezi along the Zambian border; particularly the Capui, Cuando, Luanguinga and Lueti. The grassland is short and dominated by *Loudetia simplex* and *Monocymbium ceresiiforme*. Trees are absent and are replaced by rhizomatous geoxylic suffruties which may reach 0.6 m in height. The most common suffrutex is *Parinari capensis*.

Around N'Giva, Degraded Dry Deciduous Savannah is found adjacent to the dry, open woodlands to the north, and the bushland to the south in Namibia. This is secondary vegetation dominated by stunted tree thickets on grassland, the result of the clearance of the adjacent woodlands. Clearance in this area is associated with the towns of Anhanca, Evale, N'Giva, Namancunda, Nehone and Moygua. This is a stronghold of the UNITA forces and troop concentrations have created high population pressure with rural dwellers moving into "safe" areas. Demands for agricultural land and fuelwood are high and there has also been woodland clearance for timber and military operations.

**Montane Grassland**

Montane Grassland is found on the highest parts of the Bie Plateau in the west, where the altitudes exceed 2,000 m and the soils are too thin to support the miombo woodland vegetation found in the lower parts of the plateau. It is not very extensive, covering only 833 km² or 0.01% of Angola, and is restricted to Huambo and Huila Districts.

This class has a moderately high level of productivity during the wet season from November to May. It declines rapidly in the dry season with NDVI values falling from 175 to 135 in July and October. This drop in NDVI is indicative of moisture stress due to the low water-holding capacities of thin soils, a further factor militating against tree growth.

The fuelwood potential of Montane Grassland is severely restricted by low levels of woody biomass and seasonal moisture stress.

**Coastal and Desert Vegetation**

The Namib Desert stretches up the Atlantic coast into southern Angola. It is most extensive around Namibe and Tombwa, where it extends about 100 km inland; north of Namibe it thins out. True desert is only found in the south but arid thicket vegetation is present around the fringes of the Namib Desert and along a thin coastal strip (10–15 km wide). The strip reaches as far north as N'Gunza in Kwanza Sul District (most of which is Dry Coastal Savannah and Arid Coastal Thicket). A thin band of desert-like vegetation can be seen stretching even further up the coast. This represents bare saline flats and grassy sand dunes where the only woody species is the shrubby *Strychnos spinosa*.

This biomass class is mainly restricted to Benguela, Kwanza Sul and Namibe Districts. It covers an area of 21,184 km², 1.7% of the country. It is most extensive in Namibe District (15,126 km²) where it accounts for 26.5% of the entire district. Levels of vegetation productivity are very low throughout the year with NDVI values varying only from 125 to 130.

Apart from areas of mobile sand south of Tombwa, much of the Namib Desert is grassland dominated by desert annuals, grasses, succulents and *Welwitschia bainesii*. Along the coast, where the soils are more saline, there are halophytic plant communities. Areas of gravel (especially near Namibe) and calcareous and gypsiferous inland terraces support more specialized grass flora. Woody vegetation is therefore absent from most of the desert, except for bushland developed along rivers and inland depressions. This riparian bushland is dominated by *Acacia albida*, *Ficus* spp., *Hyphaene benguellensis*, and *Tamarix usneoides*. Depressions in the eastern Namiba Desert are characterized by *Acacia albida*, *A. reficiens A. tortilis*, *Commiphora* spp., *Hoodia currori* and *Sterculia setigera*.

The fuelwood potential of the Namib Desert and surrounding areas is severely restricted. Furthermore, because of the very low annual level of productivity, any significant fuelwood exploitation will inevitably lead to an irreparable erosion of the growing stock.

**BIOMASS SUPPLY**

The estimates of growing stock and MAI (see Table 3.3) clearly illustrate the immense quantities of biomass and high annual woody-biomass yield in Angola. Both are easily the highest of any country within the SADCC region.

Consider the total estimated growing stock of 4,713 million

tonnes for the whole country in relation to Angola's population of 8.30 million (1986 estimate) which has an average density of 6.7 persons per km². It is apparent that, in per capita terms (568 t/person), biomass potential is enormous.

However, this favourable balance of woody biomass is upset by several factors. These figures disguise a considerable imbalance in woody-biomass resources when compared to the areal extent of the biomass classes. The rain forest/miombo woodland transition, although only covering 12.8% of the land area, accounts for 24.1% and 25% of the growing stock and MAI respectively. The various types of miombo woodland, which cover 51.2% of the country, have 63.1% and 62.8% of the growing stock and MAI respectively. If the degraded woodlands and savannahs (29.9% of the country) are examined the situation is even worse, as they account for only 12.5% of the growing stock and 12.8% of the MAI. Most critically, all of the arid and coastal zone and grassland vegetation, whilst accounting for 6% of Angola, only provides 0.2% of the growing stock and 0.3% of the MAI.

This analysis of growing stock and MAI data does not take into account the geographical distribution of the areas of low and high levels of biomass production. Analysis by district of the overall distribution of growing stock and MAI shows low proportions in Benguela, Kwanza Norte, and Namibe; and very high amounts in Bie, Huila, Kuando-Kubango, Lunda Norte, Malanje, Moxico and Uige (see Table 3.4). These trends are due to imbalances between population demand and woody-biomass resources; and inherent low levels of productivity.

The majority of the population inhabit the narrow coastal plain, whereas the vast wood reserves of the rain forest/Miombo Woodland transition zone and the miombo woodlands are situated in the north and on the plateaux. This has produced considerable pressure on the woody-biomass resources of the littoral and sub-littoral areas, leading to localized shortfalls of fuelwood in the coastal districts (Cabinda, Zaire, Bengo, Kwanza Sul, Benguela and Namibe).

The grassland, coastal and desert vegetation, occupying 6% of the country, contributes very little to Angola's woody-biomass resources. In fact, Namibe District suffers from a natural biomass shortage and is unable to support the urban demands generated within the district. These problems of inherent low productivity affect all of southern Angola, the coastal plain and the Zambian

**Table 3.3** ANGOLA: SUMMARY OF GROWING STOCK AND MAI DATA

| Biomass Class | (km²) | Area (% of country) | Growing Stock (mill t) | (% of total) | MAI (mill t) | (% of total) |
|---|---|---|---|---|---|---|
| Trans. Rain Forest/ Miombo Woodland | 159,680 | 12.8 | 1,137.3 | 24.1 | 35.9 | 24.9 |
| Dense High Miombo Woodland | 111,281 | 8.9 | 792.5 | 16.8 | 25.0 | 17.3 |
| Dense Medium-Height Miombo Woodland | 221,164 | 17.7 | 1,575.1 | 33.4 | 49.8 | 34.5 |
| Seasonal Miombo Woodland and Wooded Savannah | 306,946 | 24.6 | 609.3 | 12.9 | 15.0 | 10.4 |
| Dry Deciduous Savannah | 229,657 | 18.4 | 385.6 | 8.2 | 11.3 | 7.8 |
| Dry Coastal Savannah and Arid Coastal Thicket | 48,484 | 3.9 | 57.4 | 1.2 | 2.1 | 1.5 |
| Dry Inland Savannah | 26,263 | 2.1 | 31.1 | 0.7 | 1.1 | 0.8 |
| Degraded Rain Forest and Miombo Woodland | 33,220 | 2.7 | 31.4 | 0.7 | 0.8 | 0.6 |
| Degraded Dry Deciduous Savannah | 34,987 | 2.8 | 81.7 | 1.7 | 2.9 | 2.0 |
| Bushy Arid Shrubland | 15,748 | 1.3 | 11.2 | 0.2 | 0.5 | 0.3 |
| *Chanas da Borracha* Grassland | 37,251 | 3.0 | 0 | 0 | 0 | 0 |
| Montane Grassland | 833 | <0.1 | 0 | 0 | 0 | 0 |
| Coastal and Desert Vegetation | 21,184 | 1.7 | 0 | 0 | 0 | 0 |
| TOTAL | 1,246,698 | | 4,713 | | 144.4 | |

border region. These areas are dominated by savannah vegetation and occupy about 30% of the country. They contribute an estimated 561.63 million tonnes of standing biomass, or approximately 12% of the total standing biomass. Yet, because of easy access from the major concentrations of population in some of the savannah areas, there is a continual depletion of woody-biomass resources.

**Table 3.4** ANGOLA: GROWING STOCK AND MAI
DISTRIBUTION BY DISTRICT

| District | Proportion of: | |
| | Growing Stock (%) | MAI (%) |
| --- | --- | --- |
| Bengo and Luanda | 2.3 | 2.4 |
| Benguela | 1.9 | 1.8 |
| Bie | 8.1 | 8.2 |
| Cabinda | 0.3 | 0.3 |
| Cunene | 4.8 | 4.8 |
| Huambo | 3.0 | 2.8 |
| Huila | 8.4 | 8.6 |
| Kuando-Kubango | 15.8 | 16.1 |
| Kwanza N. | 1.9 | 1.8 |
| Kwanza S. | 3.4 | 3.2 |
| Lunda N. | 12.0 | 12.1 |
| Lunda S. | 5.6 | 5.3 |
| Malanje | 7.6 | 7.3 |
| Moxico | 15.2 | 15.4 |
| Namibe | 1.2 | 1.4 |
| Uige | 6.3 | 6.3 |
| Zaire | 2.3 | 2.2 |

Furthermore, extensive areas of Angola's natural forests are state-owned and forest operations are strictly regulated by the government. For example, the dry forests of central Angola yield wood which is largely processed for industrial use and it has been estimated that only 20% of the total potential yield of each of the trees found here is usable as fuelwood. Extending this figure to the whole country reveals that only approximately 20% of total annual wood production from Angola's natural forests is used for fuelwood. Even this low proportion would easily meet annual demand – potentially estimated to be 4,039,661 tonnes for the year 1990.

To add to the problem, there is restricted access to woody-biomass resources at a large scale, due to forest reservations and national parks. Information on forest reserves was unobtainable, but the distribution of national parks in 1982 shows that significantly large areas were under national park ordinances. A rough summary of the distribution is as follows:

| | |
|---|---|
| Bengo: | large areas of Dry Coastal Savannah and Arid Coastal Thicket; and, more significantly, Dense, Medium-Height Miombo Woodland (Parc Nacional da Kisama) |
| Cunene: | large areas of Dry Deciduous Savannah (Parc Nacional da Mupa) |
| Huila: | large proportions of Dense, Medium-Height Miombo Woodland (Parc Nacional da Bikuar) |
| Kuando-Kubango: | large amounts of Dense, High Miombo Woodland and Dry Deciduous Savannah (Parc de Mavinga and Parc do Luiana) |
| Malanje: | large areas of Transitional Rain Forest/ Miombo Woodland and other types of miombo woodland (Parc Nacional da Kangandala and Parc Internacional do Luando) |
| Moxico: | large areas of Dry Deciduous Savannah and Grasslands (Parc Nacional da Kameia) |
| Namibe: | most of the Desert Vegetation and Bushy Arid Shrubland areas in the Parc de Mocamedes and Parc Nacional do Iona. |

Other parks occur in Benguela and Bie Districts. Potentially the most significant of these, in terms of restricting woody-biomass exploitation on a district basis, are the parks in Namibe as they control access to almost all the woody biomass on the coastal plain. Those in Bengo have a similar potential affect because of the high fuelwood demand from Luanda. In other districts the problems are more localized because of high biomass potential.

Overriding all of the factors affecting woody-biomass supply in Angola, however, is the war situation. The effectiveness of national parks and forest reserves in restricting access to wood resources and the inequalities in woody-biomass supply potential on a district basis pale into insignificance when the restrictions that war has imposed on the Angolan rural economy are considered.

Large areas of the country outside the main towns are subject to persistent conflict, and the ability to transport woody-biomass supplies to the areas of high demand is severely limited. There is some evidence of deforestation along the routes leading into the plateaux from the coastal towns. This can be seen on the ground but not on the satellite imagery. Most importantly, the restrictions the war has placed on the Angolan transport network accentuate the problems of low woody-biomass potential in high-demand areas, such as the coastal plain, and bolsters the extremely high surplus in low demand areas.

Below, the areas of very low woody-biomass supply potential are defined by district (from Table 3.4) in an estimated order of severity:

i)   Namibe – most of the district
ii)  Bengo – especially the coastal plain (including all of Luanda District)
iii) Benguela – along the coast and inland as far as Boccoo and Cantangue
iv)  Kwanza Sul – along the coastal plain and inland as far as Gabela and Quibala
v)   Kwanza Norte – western half of the district and in the coffee plantations
vi   Zaire – along the coastal plain
vii) Cabinda – along the coastal plain
viii) Cunene – most of the south of the district, especially around N'Giva
ix)  Kuando-Kubango – in the Cuando valley
x)   Moxico – in the Luena alley, and on parts of the Zambian border
xi)  Lunda Sul – in the Luena Valley

# 4. Botswana

GENERAL
DESCRIPTION AND
BIOMASS CLASSES

Botswanan biomass resources strongly reflect the two main constraints on vegetation growth in the country: the sparsity and seasonality of rainfall; and the low moisture-holding capacities of sandy soils. The vegetation types in the country range from restricted occurrences of highly productive riparian woodlands to bare salt pans. Areally, the greatest proportion of biomass resources lie in the woodlands, bushlands and shrublands. Dense closed-canopy woodlands, which as water resources become scarcer take on a more open nature, are found in the north. Lanly (1981) suggests such woodlands are not areally significant in Botswana, but this study suggests they account for 6.8% of the country. As water becomes scarcer still, there is an increase in the proportion of shrubs, and woodlands grade into bushland and shrubland.

On a national scale, woody-biomass resources are quite high when compared to the total population. However, because the population is restricted to the east there are substantial regional disparities in biomass-resource supply. The crisis area is undoubtedly the densely populated eastern sector. Although the pattern of vegetation is strongly influenced by available water resources, the complex mosaic of woodland degrading into bushland is strongly indicative of vegetation clearance for agriculture, mining, communications and settlements. Less densely populated areas, such as the extreme north and the Kalahari Desert, have a biomass-resource surplus.

Nevertheless, with the exception of the extreme north of the country, vegetation productivity levels are generally low throughout Botswana. Therefore, although standing biomass in many regions may fulfil present fuelwood requirements, the ability of the vegetation to recover from large-scale exploitation, or to produce a sustained annual fuelwood crop, is more restricted.

Low productivity levels and low standing biomass are typical of large areas of the following districts:

- Barolong
- Ghanzi
- Kgalagadi
- Kgatleng
- Kweneng
- Ngwaketse
- South East

Of these districts the following are either highly populated, or have large towns (shown in parentheses):

- Kgatleng (Mochudi)
- Kweneng (Molepolole)
- Ngwaktse (Kanye)
- South East (Gaborone, Lobatse)

In addition the following large towns, which are all in Central District and have populations larger than 10,000, are situated in biomass classes with localized supply problems:

- Mahalapye
- Serowe
- Selebi Phikwe

Only two large Botswanan towns are in biomass classes with less severe problems:

- Francistown
- Maun

Botswana is divided into nine biomass classes (see map on p.57). This division is based on the interpretation of NOAA-7 AVHRR GAC data and reference to previous botanical and forestry studies of the country. The biomass classes are:

Dense Woodland
Open Woodland
Woodland and Bushland
Riparian Woodland
Fringing Palm Grassland and Woodland
Hill Woodland and Shrubland
Bushland with Scrubby Woodland and Woody Shrubland

**Key to facing page:**

B Riparian Forest
C Dense Woodland
I Open Woodland
K Woodland and Bushland
L Bushland with Scrubby Woodland and Woody Shrubland
M Shrubland and Bushy Shrubland
N Hill Shrubland and Woodland
O Fringing Palm Woodland
P Salt Pans

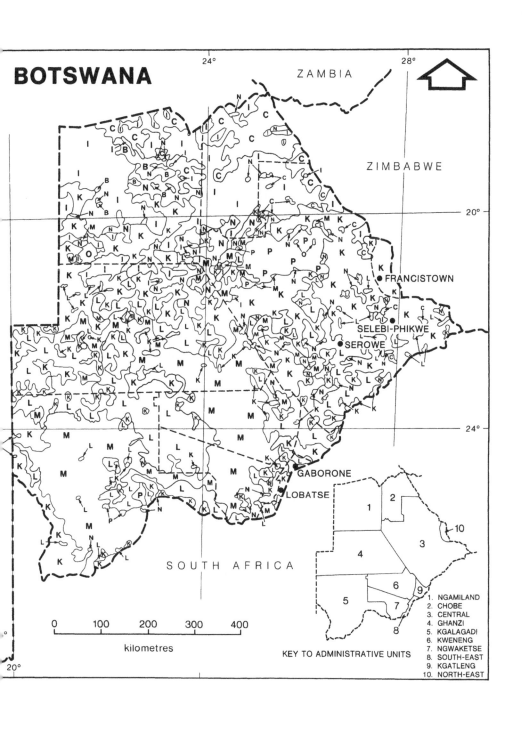

# BOTSWANA

ZAMBIA

ZIMBABWE

24°

28°

20°

FRANCISTOWN

SELEBI-PHIKWE

SEROWE

24°

GABORONE

LOBATSE

SOUTH AFRICA

20°

0    100    200    300    400

kilometres

KEY TO ADMINISTRATIVE UNITS

1. NGAMILAND
2. CHOBE
3. CENTRAL
4. GHANZI
5. KGALAGADI
6. KWENENG
7. NGWAKETSE
8. SOUTH-EAST
9. KGATLENG
10. NORTH-EAST

Shrubland and Bushy Shrubland
Salt Pans

The Salt Pans class will not be considered further due to the absence of woody vegetation.

## DESCRIPTIONS OF INDIVIDUAL BIOMASS CLASSES

### Dense Woodland

Dense Woodland is restricted to the extreme north of Botswana; it is particularly important in Chobe and Ngamiland, and to a lesser extent in Central District. It has an areal extent of 33,361 km² or 5.7% of the country. It is characterized by a well-developed tree canopy and a thicket-like shrubby under-storey. It is floristically diverse and has been divided into a number of vegetation types by previous workers. Biomass productivity is characterized by a moderately marked seasonality.

There is a period of high productivity (NDVI ranging from 170–180) between January and April. This is followed by a gradual decrease in productivity from the high levels in April to lower values (NDVI approximately 150) in September. This is a response to increasing moisture stress. Productivity levels rise again towards the end of the year.

In the Chobe area, Dense Woodland has been classified as "Dry Deciduous Forest" (Weare & Yalala, 1971) with *Baikiaea plurijuga* dominating the vegetation on the areas of deeper Kalahari Sands. The *B. plurijuga* woodland canopy is usually about 20 m high and is taller than the more open woodlands to the south. Associated with *B. plurijuga* on deep, sandy substrates are *Ricinodendron rautanenii, Pterocarpus antunesii, Entandrophragma caudatum, Burkea africana* and *Erythrophleum africanum*. The invasive *Acacia erioloba* and *Combretum collinum* are also widespread.

There is no well-defined lower canopy (White, 1983) but several trees are subdominant. The shrub layer forms a thicket of tall (5–8 m) coppicing shrubs, the most common species being *Baphia obovata, Popowia obovata, Bauhinia petersiana* (including *B. macrantha), Combretum celastroides, C. elaeagnoides* and *Acacia ataxacantha*. Where the Kalahari Sands are shallower, *Lonchocarpus capassa, Terminalia sericea* and *Ochna pulchra* enter the community and *Burkea africana* becomes the dominant species. Further west *Brachystegia boehmii* woodland is found. This consists mainly of *B. boehmii, Isoberlinia globiflora* and *Swartzia madagascarensis*.

Around the Okavango Delta the woodland has been classed as

"Ngamiland Tree Savannah" (Weave & Yalala, 1971) and consists mainly of *Acacia erioloba, Combretum imberbe, C. mossambicense, Terminalia sericea, Zizyphus mucronata* and *Acacia mellifera*. The shrub layer here consists of shrubby *Colophospermum mopane, Grewia flava* and *Croton megalobotrys*.

Dense Woodland provides high growing stocks and generally high levels of productivity which show a slight seasonality. Dense Woodland is relatively unexploited and could withstand controlled fuelwood exploitation.

**Open Woodland**

Open Woodland forms a broad belt across northern Botswana with important outlying regions around Nata and the Ghanzi Ridge. It is particularly important in Central, Chobe, Ghanzi and Ngamiland Districts and has an areal extent of 83,437 km², or 14.3% of the country. Floristically, it is very similar to Dense Woodland but there are major differences in the structure of the vegetation, particularly in the more open canopy woodland. Around the Ghanzi Ridge, this woodland is open enough for shrubs to invade and can therefore be classed as scrubby woodland. Phenologically it is quite distinct from Dense Woodland, although similar in terms of productivity and biomass. It is characterized by a long, gradual decline from high levels of productivity in January (NDVI of about 180) to low levels of productivity in September and October (NDVI approximately 140). Productivity increases sharply again at the end of the year.

The Open Woodland in the north is composed chiefly of *Acacia* spp., *Combretum* spp., *Terminalia prunioides* and *Zizyphus mucronata* with shrub *Colophospermum mopane, Grewia flava* and *T. sericea*. In the northwest, scrub woodland occurs along the quartzite- and limestone-rich Ghanzi Ridge where the vegetation is denser than in the surrounding areas of bushland and shrubland. Scrub woodland is chiefly distinguished from bushland and shrubland to the south by its greater overall density of trees, and by the addition of *Colophospermum mopane* to the flora. There are similar densities of shrubs in the bushland and woodland areas but in the shrubland areas, the densities are greater. Grasses are relatively unimportant here but shrubs play a major role. Trees found here include *Combretum* spp., *Albizia anthelmintica, Boscia rehmanniana, Acacia* spp., *Lonchocarpus nelsii, Peltophorum africanum* and *Montinia caryophyllacea*.

There is a distinct grassland transition zone between Dense and

Open Woodland around the Mabebe Depression, which is dominated by *Cenchrus ciliaris* and *Chloris gayana*.

Fuelwood resources in the Open Woodland are significantly lower than in the Dense Woodland. This is due to the lower growing stock and lower levels of annual productivity. Nevertheless, it does provide a potential fuelwood source in northern Botswana as it covers 14% of the country and extends into the populous north-east. Fuelwood exploitation would need to be carefully monitored and controlled due to the low annual productivity.

**Woodland and Bushland**

This biomass class occurs in northern and eastern Botswana to the south of the Open Woodland and is the largest biomass class in Botswana. It forms a strip extending from the South Africa–Zimbabwe border, south of the Makgadikgadi Pan, to the Namibian border. There is also a narrow strip to the north and north-east of the Makgadikgadi Pan and a number of outlying regions to the south of the pan. It is mainly found in Central, Ghanzi, Kgalagadi, Kgatleng, Kweneng, Ngwaktse, and North-East Districts and has an areal extent of 197,664 km² covering 34% of the country.

This woodland-bushland transitional zone exhibits lower levels of productivity during the wet season than the other woodland types. Consequently, it has a less marked seasonality. During the wet season the NDVI values are low, ranging from around 150 to 155, with peaks of around 160 in April. This is followed by a long slow decline into September and October when the lowest NDVI values, of about 140, are found. Productivity levels increase sharply after this.

In the east, this biomass class is ecologically similar to the Open Woodland found to the north, and is dominated by *Colophospermum mopane*. However there is an ecological gradient to the west and, once it begins to meet the sandier soils developed on the Kalahari Sands, it grades into bushland and shrubby bushland.

Floristically and structurally, this class is quite diverse and corresponds closely to other vegetation types. It is differentiated from the adjacent biomass classes by its phenological characteristics. Although the growing stock appears to be relatively high when compared to the adjacent vegetation to the south, the annual level of productivity is low. This limits the amount of fuelwood that can be extracted from the Woodland and Bushland biomass class.

**Riparian Woodland**

Riparian Woodland is locally important in the Okavango Delta and Chobe Valley on the Namibian border in Ngamiland District. It has an areal extent of 6,297 km² or 1.1% of the country. It is typically a fringing forest with floristic similarities to the surrounding woodland types, as well as having species adapted to the edaphic conditions. It exhibits the highest levels of productivity of any biomass class in Botswana but this is characterized by seasonal variations. Productivity is highest in April when NDVI values are about 185; this declines slowly until a period of low productivity is reached in September and October (NDVI of about 160). Levels then increase, initially quite sharply and then more slowly, until the following April.

Around the Okavango Delta, swamp grassland is fringed by a belt of large trees such as *Ficus sycomorus, Diospyros mespiliformis, Ionchocarpus capassa, Combretum imberbe, C. transvaalense, Colophospermum mopane, Acacia erioloba, Croton megalobotrys, Acacia xanthopholea, A. galpinii* and *Hyphaene crinata*. Levels of production and woody biomass are high in the Riparian Woodland but there are serious accessibility problems:

i) much of the wood is in the tree crowns;
ii) parts of the Riparian Woodland are flooded for much of the year; *and*
iii) almost one-third of this biomass class is found in the Chobe and Okavango National Parks.

**Fringing Palm Grassland and Woodland**

Palm Woodland is locally important at the eastern end of Makgadikgadi Pan in Central District and has an areal extent of only 3,115 km², just 0.5% of the country. It is characterized by low levels of productivity with little seasonal variation; nevertheless, productivity levels are higher than on the surrounding salt pan. Very low productivity (NDVI of about 125 to 130) characterizes the vegetation between August and February. After February, levels increase and are highest between April and July, when the NDVI reaches 140.

The slightly higher ground to the east of Makgadikgadi Pan can be considered as a transition zone between the fringing pan-grassland around the pan itself, and the Woodland and Bushland further east. There are large areas of flat, open grassland which support few trees. The main grass species here are *Aristida* spp.,

*Sporobolus spicatus,* and *Odyssea paucinervis.* This grassland is interspersed in places with belts of trees such as *Colophospermum mopane, Combretum imberbe* and *Acacia nigrescens. A. kirkii* and *A. nilotica* trees are found widely scattered throughout the grassland area whilst on drier, raised hummocky areas clumps of *Adansonia digitata, Terminalia* spp., *Albizia* spp., and palms of *Hyphaene* spp. are found. In some areas *Hyphaene* palms become more common and reach approximately 8 m in height.

Throughout the area, the shrub layer is poorly developed and the overall structure of the community can be summarized as grassland supporting a scattered tree cover which increases in localized areas.

The low productivity and biomass of this class, combined with its inaccessibility, means that it is not exploited at the present time or in the near future.

## Hill Shrubland and Woodland

Hill Shrubland and Woodland is locally important in eastern and northern Botswana (Central, Chobe, Ghanzi, Kgalagadi, Ngamiland and Ngwatske Districts) and has an areal extent of 26,502 km² or 4.6% of the country.

It is found on small hills reaching heights of 100 m in the east of the country such as the Lobatse Hills, Mokgware Hills, Tswapong Hills and the hills around Francistown. It is an open woodland and shrubland vegetation that differs floristically to that on lower ground. Phenologically, it shows certain similarities to the woodland categories but it has lower wet-season levels of productivity. However, these levels are higher than the shrubland and bushland found on lower ground to the south. Productivity is highest in January (when NDVI is about 175) and declines until May when a slight increase in productivity is noticeable. After May, there is a sharp decline in productivity into the dry season with the lowest productivity (NDVI 140) occurring in July. It remains low throughout the dry season until October and then increases sharply again.

It is generally dominated by *Colophospermum mopane, Lonchocarpus* spp., *Cassia* spp., *Ficus sycomorus, F. capensis* and *Adansonia digitata.* However the Shoshong and Mokgware Hills, which are covered by open to fairly dense scrub woodland, are dominated by *Acacia nigrescens* together with *Combretum apiculatum, Sclerocarya caffra* and *Croton gratissimus.* Other species found include *Pappea capensis, Combretum molle, Spirostachys africana, Euphorbia ingens* and *Albizia harveyi.*

Fuelwood supplies from Hill Shrubland and Woodland are

limited. First, the growing stock is low because of previous exploitation. Secondly, the thin soils and highly seasonal production mean that annual productivity is limited. It can only be considered in a local context for fuelwood supply because of its scattered distribution.

**Bushland with Scrubby Woodland and Bushy Shrubland**

This biomass class is most strongly developed along the South African border between Gaborone and Francistown, in the central Kalahari Desert and to the south-west of Makgadikgadi Pan. It is found in all districts except Chobe and North-East and has an areal extent of 106,743 km² or 18.4% of the country. The eastern areas are dominated by Bushland with patches of Scrubby Woodland and Bushy Shrubland. This complex of different vegetation types are categorized by ecological structure rather than flora and are accentuated by clearance around settlements. To the west, Bushland is still dominant but there is proportionately more Bushy Shrubland and little Shrubby Woodland. Consequently, it is floristically diverse and has been divided into a number of vegetation types by previous workers.

Phenologically, this biomass class is very similar to the areas of Woodland and Bushland that lie to the north of it. The main difference is its lower annual level of productivity. This is relatively low in the wet season, with NDVI values ranging from about 150 to 155 up to peaks of about 160 in April. This is followed by a long, slow decline in productivity until the lowest values (NDVI of about 140) are found in September and October. Productivity increases sharply after this.

The bushland along the eastern side of the country is composed of small- to medium-height trees with a well-developed shrub layer. In the south, the dominant tree is *Peltophorum africanum* with *Terminalia serica*, *Burkea africana* and *Boscia albitruncata*. The main shrubs are *Dichrostachys cinerea*, *Acacia* spp., *Bauhinia macrantha* and some *Ochna pulchra*. Further north the bushland is dominated by scattered trees of *Acacia* spp., *Sclerocarya caffra*, *Peltophorum africanum*, and *Combretum imberbe*. Amongst these trees are numerous smaller trees such as *Combretum apiculatum*, *Terminalia sericea*, *Acacia* spp., *Boscia albitruncata*, *Zizyphus mucronata*, *Albizia* spp. and *Commiphora schimperi*. The main shrubs here are *Dichrostachys cinerea*, *Grewia flava* and *Grewia* spp.

To the north of the Tswapong Hills, the vegetation changes both floristically and structurally as *Colophospermum mopane* becomes important. This species occurs both as a medium-height tree and in a shrub form, and is usually the dominant species among the vegetation in which it occurs. Associated trees include *Acacia nigrescens, Sclerocarya caffra, Terminalia prunioides, Commiphora mossambicensis, Combretum* spp. and *Burkea africana*. The most commonly found shrubs are *Acacia* spp., *Dichrostachys cinerea, Grewia* spp. and *Commiphora pyracanthoides*.

The Scrub Woodland to the north of the Tswapong Hills represents the vegetation with the largest biomass potential within this class. It has floristic similarities to the Open Woodland and is dominated by *Colophospermum mopane*. The vegetation composition is similar to that of the mopane woodland and bushland described in the Woodland and Bushland biomass class. The associated trees have variable geographical distributions and may become locally dominant. They include *Acacia nigrescens, Combretum* spp., *Kirkia acuminata, Sclerocarya caffra, Commiphora mossambicensis* and *Burkea africana*. The first two species are the most common over large areas. The shrub storey consists of varying proportions of *C. mopane, Acacia* spp., *Terminalia prunioides, Dichrostachys cinerea, Zizyphus mucronata* and *Grewia* spp.

The eastern bushland differs from that found in the Kalahari Desert and the north-west of the country in that the shrub layer is better developed and the density of trees is much higher. However, the canopy is still generally fairly open and, especially in the areas dominated by *Colophospermum mopane*, the grass cover is usually poor. The trees are usually less than 12 m high, with average heights ranging between 7 and 9 m.

This biomass class already shows extensive evidence of clearance for fuelwood and in many areas it represents a degraded form of the better woodlands and bushlands. The growing stock is low and the productivity is limited by the seasonality in production and the negative feedback effects of declining soil fertility. It is being over-exploited at the present time at rates which exceed annual productivity. Levels of productivity will continue to decline in the near future, placing even greater limitations on supplies of fuelwood from this area.

**Shrubland and Bushy Shrubland**

Shrubland is mainly restricted to the Kalahari Desert and is mainly found in Kgalagadi, Kweneng and Ngwakste Districts with lower amounts in Barolong, Central, Ghanzi, Kgatleng and South-East Districts. It covers some 17.9% of the country, or 104,903 km². It has floristic affinities with the surrounding vegetation types – bushland and scrub woodland – but is structurally quite distinct. In the Kalahari, the trees are always less than 7 m tall, and often much shorter; normally, they are widely spaced. The vegetation consists of relatively few species which appear in various combinations to form mosaics which are usually related to topographical features (ridges, pans, dune crests and troughs). The difference between the two categories in this class reflect differences in overall vegetation density and in the relative proportion of trees present.

It is also quite distinct phenologically, with relatively low levels of productivity throughout the year: the NDVI is always below 150. Productivity is highest from January to May when NDVI varies between 140 and 150; it then declines slowly until the lowest levels are reached in October and November, when the NDVI has declined to 135.

Three sub-types of Kalahari vegetation can be recognized, reflecting the broad increase in rainfall from less than 250 mm per annum in the extreme south-west, to between 400 and 450 mm in the north-east:

   i)  On the rolling sand dunes in the south-west, the vegetation is very sparse with few, widely scattered trees of *Boscia albitrunca* and *Acacia mellifera (detinens)* occurring mainly on the dune crests. Shrubs such as *Rhigozum trichotomum* are confined to the troughs between the dunes. Shrubs which are confined to this south-western area include *Acacia haematoxylon* and *Monechma* spp.

   ii) North-east of this area, the vegetation is classed as "Southern Kalahari Bush Savannah" (Weare & Yalala, 1971). Again, it occurs in rolling, sandy country with wide plains, depressions and pans. The main tree species are *Acacia erioloba, A. mellifera (detinens)* and *Boscia albitrunca*. Low shrubs include *Grewia* spp., *Boscia albitrunca, Acacia hebeclada, Dichrostachys cinerea, Bauhinia macrantha* and *Zizyphus mucronata*. Generally, the densest growth of trees and shrubs are on the rises; the depressions are more open.

   iii) The northern part of the Kalahari is a tree-and-bush savannah reflecting the slightly larger tree component in the vegetation. Here again, tall trees are mostly confined to the sand ridges and include *Burkea africana, Peltophorum africanum, Terminalia sericea, Croton zambesicus, Lonchocarpus nelsii, Rhus tenuinervis, Combretum* spp., *Acacia* spp., *Boscia albitrunca, Commiphora* spp. and *Ochna pulchra*. Low-growing shrubs are common on the plains and between the taller trees; these are mainly *Croton subgrattissimus, Grewia* spp., *Ximenia caffra* and *Commiphora pyracanthoides*.

Although the growing stock appears to be quite high in this biomass class, the annual productivity is very low. This latter factor seriously limits the level of extraction from this biomass class and it is certain that any but the smallest rates of fuelwood extraction would exceed the annual rates of production.

**BIOMASS SUPPLY**

Biomass resources in Botswana are very badly distributed with respect to the population. Comparisons of the total growing stock (1311.1 million tonnes) and MAI (45.6 million tonnes) with the total population of 978,000 (1986 estimate) – giving fuelwood ratios of 1334 million tonnes growing stock per person and 46.10 tonnes MAI per person – although high, are of little use. This is because the biomass classes with the highest growing stock and MAI levels are located in the far north, whilst the majority of the population live in the easternmost third of the country. This disparity can be seen in Tables 4.1 and 4.2.

The three woodland classes – Dense, Open and Riparian – cover 123,095 km² or 21.2% of the country. Yet their proportions of the growing stock (44.9%) and MAI (39%) exceed the areal proportion of these classes (see Table 4.1). Of these, only Open Woodland occurs to any great extent in the highly populated eastern sector.

The populated zone is dominated by three biomass classes: Woodland and Bushland; Bushland with Scrub Woodland and Bushy Scrubland; and Hill Woodland and Shrubland. Only in one of these classes – Woodland & Bushland – does the proportion of growing stock (35.4%) and MAI (36.4%) exceed the areal proportion of 34% (see Table 4.1). This is not surprising as all of these classes represent to a certain extent degraded and less productive forms of woodland.

As a consequence, Chobe and Ngamiland Districts, in the far north, hold between them 40.9% of the growing stock and 37.1% of the MAI (see Table 4.2). Urban demand in these districts is only

**Table 4.1**  BOTSWANA: SUMMARY OF GROWING STOCK AND MAI DATA

| Biomass Class | Area (km²) | Area (% of country) | Growing Stock (mill t) | Growing Stock (% of total) | MAI (mill t) | MAI (% of total) |
|---|---|---|---|---|---|---|
| Dense Woodland | 33,361 | 5.7 | 237.6 | 18.2 | 7.5 | 16.7 |
| Open Woodland | 83,437 | 14.3 | 308.5 | 23.2 | 9.1 | 19.2 |
| Riparian Woodland | 6,297 | 1.1 | 44.9 | 3.4 | 1.4 | 3.1 |
| Woodland and Bushland | 197,664 | 34.0 | 461.8 | 35.4 | 16.4 | 36.4 |
| Fringing Palm Woodland | 3,115 | 0.5 | 2.2 | 0.2 | 0.1 | 0.2 |
| Bushland with Scrub Woodland and Bushy Shrubland | 106,742 | 18.4 | 151.8 | 11.6 | 6.4 | 14.2 |
| Scrubland and Bushy Scrubland | 104,093 | 17.9 | 74.0 | 5.7 | 3.2 | 7.1 |
| Hill Woodland and Shrubland | 26,502 | 4.6 | 30.3 | 2.3 | 1.5 | 3.2 |
| Salt Pans | 20,584 | 3.5 | 0 | 0 | 0 | 0 |
| TOTAL | 581,795 | | 1,311.1 | | 45.6 | |

**Table 4.2**  BOTSWANA: GROWING STOCK AND MAI DISTRIBUTION BY DISTRICT

| District | Proportion of Growing Stock (%) | MAI (%) |
|---|---|---|
| Barolong and SE | 0.2 | 0.2 |
| Central | 23.0 | 23.5 |
| Chobe | 8.1 | 7.2 |
| Ghanzi | 18.8 | 19.6 |
| Kgalagadi | 10.0 | 11.4 |
| Kgatleng | 0.6 | 0.7 |
| Kweneng | 3.1 | 3.6 |
| Ngamiland | 32.8 | 29.9 |
| North East | 1.0 | 1.0 |
| Ngwaketse | 2.5 | 2.9 |

generated by one major town, Maun. In contract, Barolong and South-East Districts, around Gaborone, have a very high urban demand but only 0.2% of the growing stock and MAI.

In the eastern sector there are five further factors which affect the biomass supply figures, apart from the low levels of standing stock and yield:

i) Strong firewood preferences amongst the population have been identified by Jelenic & van Vegten (1981), (see Table 4.3). They emphasize the following required qualities of preferred fuelwood tree species: low-density wood, so that large loads can be carried; ease of breakdown into small pieces; easy ignition; hot, long-lasting fire; and low noxious smoke capacity.

The collection of the most preferred species (e.g. *Combretum apiculatum* and *C. imberbe*) occurs first. Then alternative woods with lower rates of preference are collected (e.g., *Acacia erubescens* and *Dicrostachys cinerea*). Indicators of selective removal pressure can be seen when low-preference trees, such as the protected *Boscia* spp., are cut; and when woody-tree vegetation has reverted to thorn bush.

ii) The recurrent droughts that occur in southern Africa depress the MAI of all the woody vegetation types. Paradoxically, tree mortality, by both direct and indirect causes, increases the amount of dead wood available for fuel. However, this is only a short-term benefit in the readily accessible woody-biomass resource, gained at the expense of decreased yield in future years, that can only be replaced by regeneration.

iii) Annual burning to improve forage quality for cattle is widespread. This seriously affects regeneration at the seedling and sapling stages of most species. Nevertheless, many species of *Acacia* require fire to promote seed germination, though at the mature stage, fire-resistance is variable.

iv) Conflicting wood-uses which affect fuelwood supply are evident in eastern Botswana. The most important of these are the Drought Relief Destumping Project and the demand for poles.

v) As cattle rearing is important throughout Botswana, the affects of overgrazing on the woody-biomass production system are critical to woody-biomass supply estimates. Overgrazing leads to a decreased grass and herb cover, increasing water run-off and soil erosion by water and wind. Soil moisture budgets are seriously imparied, increasing the severe dry-season moisture stresses and depressing the MAI. With the reduction in forage resources, browsing becomes more important and further depresses the MAI by preferential browsing on new shoots. The effects of browsing and trampling together work against the efficient regeneration of tree seedlings, and provide an advantage to competitive shrub species. This ultimately leads to a greater shrub component and a change from woodland to woodland and shrubland.

In western Botswana, woody-biomass resources are dominated

**Table 4.3** PREFERRED FUELWOOD TREES IN EASTERN BOTSWANA

| Species | Tree density range (trees/ha) |
|---|---|
| *Acacia erioloba* | * |
| *A. erubescens* | 20–110 |
| *A. fleckii* | * |
| *A. leuderitzii* | * |
| *A. mellifera* | 40–70 |
| *A. nigrescens* | * |
| *A. tortilis* | 30–130 |
| *Boscia albitruncata* | 2–10 |
| *B. foetida* | 3–12 |
| *Combretum apiculatum* | 10–30 |
| *C. erytrophyllum* | 6–15 |
| *C. hereroense* | * |
| *C. imberbe* | * |
| *C. zeyheri* | * |
| *Colophospermum mopane* | * |
| *Dichrostachys cinerea* | 10–50 |
| *Peltophosum africanum* | 1–25 |
| *Terminalia sericea* | 30–193 |
| *Ziziphus mucronata* | * |

Adapted from Jelenic and van Vegten (1981)     * = no data

by the Shrubland and Bushy Shrubland, and Bushland with Scrubby Woodland and Woody Shrubland, biomass classes. Although these classes cover a little over a third of the country (36.8%), on a national basis, they have disproportionately low levels of growing stocks (17.3%) and MAI (21.3%). Fortunately this area is, at present, underpopulated and the demand for woody-biomass fuel is low. White's (1979) estimates show that the woody-biomass proportion is very variable and the firewood component (defined as dead wood), varies from 1.3% in the less woody grassland and shrublands to 12% in the *Acacia* and *Terminalia* woodlands.

**Table 4.4** PROPORTION OF STANDING VOLUME AND GROWTH RATES IN DIFFERENT ECOLOGICAL AND FUNCTIONAL CLASSES IN WESTERN BOTSWANA

| Vegetation Type | Trees | Proportion (%) Shrubs | Firewood | Total (m³/ha) |
|---|---|---|---|---|
| *Acacia* woodland | 74.7 | 25.3 | 12.0 | 14.67 |
| Open Grass and Shrublands | 14.1 | 85.9 | 1.3 | 7.74 |

Source: White (1979)

The trend towards settlement schemes and ranching in the Kalahari has led to noticeable "rings of deforestation" around the main villages such as Hukuntse, Kang, Lehututu, Lokgwabe, Tshanbong and Tshare.

On a regional basis, the biomass resources in northern Botswana (Chobe and Ngamiland Districts) far exceed demand. In the east, there is great pressure on woody-biomass resources from multiple end-uses. It is likely that, overall, supply does not meet demand but there are pockets, such as the Hill Woodland and Shrubland areas, in which resources exceed current demands. In western and central Botswana there is a restricted resource base, both in terms of growing stock and productivity, but one which in most places still exceeds demand.

Therefore, Botswana can be divided into three zones of wood-supply potential.

1. Areas with high levels of growing stock and productivity, and low local demand: Chobe and northern Ngamiland Districts.
2. Areas with moderate levels of existing growing stock, restricted productivity and low local demand: south-western Central, Ghanzi, Kgalagadi, western Kweneng, southern Ngamiland and Ngwaketse Districts.
3. Areas with a generally restricted supply base, both in terms of growing stock and productivity, and a high local demand: Barolong, eastern Central, Kgatleng, eastern Kweneng, eastern Ngwaketse and North East Districts.

Zone 1 is resilient and could stand relatively high rates of exploitation. However, the resource base in Zone 2 is limited by already degraded woody-biomass stocks and low levels of productivity; and the potential in Zone 3 is equally restricted by low levels of productivity.

# 5.    Lesotho

GENERAL
DESCRIPTION AND
BIOMASS CLASSES

Lesotho is an extremely rugged, mountainous country set high in the Drakensberg. Its vegetation is totally unlike any other SADCC member state because of its high elevation and southerly location. Studies from the adjacent Natal Drakensberg to the north, suggest that the vegetation in Lesotho should fall into three altitudinal belts. In these belts the natural vegetation should be:

   i)   a Montane belt between 1,280–1,830 masl with a natural vegetation of *Podocarpus* Forest;
   ii)  a Sub-Alpine belt between 1,830–2,865 masl with a natural bushland vegetation; *and*
   iii) an Alpine belt above 2,865 masl with alpine grassland and heathland communities.

It is generally agreed that the upper limit of woody vegetation in Lesotho is between 2,400 and 2,600 masl (Herbst and Roberts, 1974; Richardson, 1983). The species suggested for woodlot cultivation in the country are various *Pinus* spp., *Cupressus glabra, Salix vaminalis, Eucalyptus macarthurii, Populus canescens, Robinia pseudoacacia* and *Acacia* spp. (Richardson, 1983).

The four biomass classes identified in Lesotho are dominated by low-grass and herb communities; woody vegetation is uncommon. At higher altitudes there are low, woody-heath communities. In the western plains, escarpment zone, and river valleys there are patches of scrubby bushland, scrub woodland and riparian woodland. Woody vegetation only takes the form of bushes and stunted trees, either as relicts of the original woodland or as a result of the invasion of degraded rangeland. Consequently, woody vegetation is very scarce and is at a premium as a fuelwood resource.

The absence of woody vegetation has been brought about by a number of factors:

i) much of Lesotho suffers from a severe montane climate which is not conducive to tree growth and means that the Alpine and Sub-Alpine Belts are predominantly grassland or heathland;

ii) this problem is exacerbated by a deeply dissected topography and associated thin soils which cannot support dense woody vegetation;

iii) the climate is milder in the valleys and on the highveld to the west, and in these areas environmental factors such as soils and available moisture are also more favourable for tree growth; *and*

iv) most woody vegetation has now been destroyed due to land clearance (for cultivation, grazing and fuelwood) leading to extremely serious soil erosion problems throughout the country and a decline in soil fertility that slows the regeneration of woodland stands.

It can be concluded from the above that Lesotho's environmental problems are due in part to the over-exploitation of natural resources. Both land clearance and over-exploitation have become increasingly prevalent since the last half of the nineteenth century with the Sotho becoming increasingly restricted to the mountains as the settlers colonized the more fertile plains to the west. Similar woody vegetation to that which once covered parts of Lesotho can be clearly seen on the remotely-sensed imagery in the South African national parks to the north and west of the country. The boundaries of these high biomass and productivity classes are so marked that the northern and western borders of Lesotho can almost be mapped from them.

The absence of woodland is reflected in the fact that in 1981, 60% of non-commercial energy consumption was derived from shrubby-bush vegetation and 40% from animal dung (DoE, 1985). None came from woodlands. Given the low growing-stock and annual increment, this suggests serious over-exploitation of the few remaining woody-biomass resources. Frolich (1984) estimated the total biomass (tree, shrub, bush and grass) resources in 1980 to be 9.17 million GJ, or 60.85% of the total energy resources. This estimate belies the severe problems of accessibility to many of the alpine and subalpine grassland and heathland resources, and the low annual levels of productivity found in most of the country.

**Key to facing page:**

R Escarpment Grassland with Scrub Woodland
S Highveld and Riparian Grassland
T Alpine/Sub-Alpine Grassland and Heathland
W Escarpment and Riparian Woodland

The assessment of the woody-biomass resource potential of Lesotho in this survey was hampered by a technical problem. Most areas of woodland are either found in small family-owned woodlots, along streams or in ravines, or as small patches of *shallahalla* bushland. In each case the areas concerned are usually less than the 4 x 4 km pixel size used in the survey. Therefore, the areas of higher biomass and productivity do not represent large areas of woodland or shrubland, but rather grassland in which the proportion of woodland and shrubland is relatively high. Only with finer resolution remotely-sensed data, could individual areas of woodland and shrubland be identified.

All of the 10 districts in Lesotho have serious environmental problems of which scarcity of woody-biomass resources are an integral part. Areas with relatively higher levels of biomass resources (mainly *shallahalla* bushland and stunted woodland) are:

i)    along the escarpments of the Front and Thaba-Putsoa Ranges in the Districts of Berea, Butha-Buthe, Leribe, Maseru and Mohale's Hoek; *and*

ii)   along the valleys of the Makhalaneng, Mohokare and Sengu (Orange) Rivers in the Districts of Mafeteng, Maseru, Mohale's Hoek, Qacha's Nek, Quthing, Thaba-Tseka.

It should be pointed out however, that in any one of these districts, woody vegetation rarely exceeds 5% of the total area.

Lesotho has been divided into four biomass classes (see map on p.75) on the basis of NOAA-7 GAC data and previous ecological and rangeland studies (Guillarmod, 1968; White, 1983; Werger 1983). These classes are:

Escarpment and Riparian Woodland
Escarpment Grassland with Scrub Vegetation
Highveld and Riparian Grassland
Alpine/Sub-alpine Grassland and Heathland

## DESCRIPTIONS OF INDIVIDUAL BIOMASS CLASSES

### Escarpment and Riparian Woodland

Escarpment Woodland is found in a fragmented north-east to south-west trending strip corresponding to the Western Escarpment of Lesotho. This Escarpment consists of two mountain ranges, the Front and Thaba-Putsoa Ranges, and overlooks the

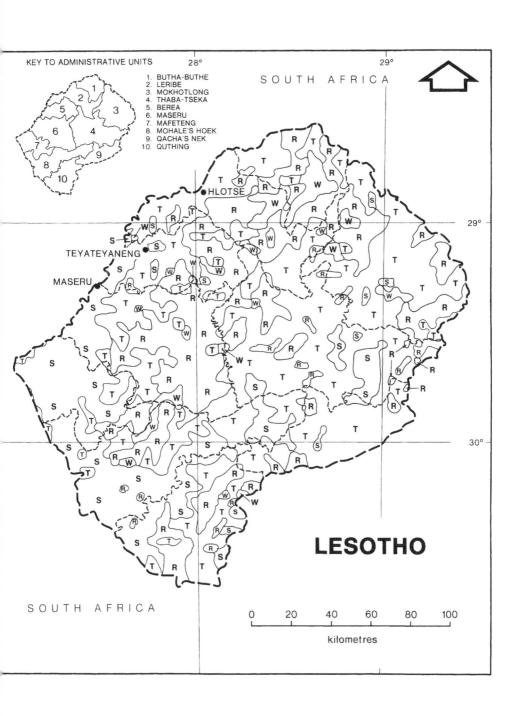

KEY TO ADMINISTRATIVE UNITS 28°

1. BUTHA-BUTHE
2. LERIBE
3. MOKHOTLONG
4. THABA-TSEKA
5. BEREA
6. MASERU
7. MAFETENG
8. MOHALE'S HOEK
9. QACHA'S NEK
10. QUTHING

SOUTH AFRICA

HLOTSE

TEYATEYANENG

MASERU

LESOTHO

SOUTH AFRICA

0    20    40    60    80    100

kilometres

Highveld of western Lesotho and the Orange Free State. It has an annual rainfall of 800–1,000 mm, and thin, stony soils. The dominant vegetation in this biomass class is bushy woodland occurring as isolated patches within larger areas of grassland and scrubby vegetation (Escarpment Grassland with Scrub Woodland). It is also one of the two main cultivated areas of Lesotho. Escarpment Woodland occurs in Berea, Butha-Buthe, Leribe, Maseru and Mohale's Hoek Districts.

Riparian Woodland occurs along most of the large rivers. The few remaining areas with high growing stock and high levels of productivity are classified in this biomass class. They occur along the Sengu (Orange) and Matsoku Rivers in Leribe, Quthing and Thaba-Tseka Districts. These valleys are the other main areas of cultivation. In all, this class covers 1,371 km², or 4.5% of the country. Most of the areas designated as forest land (3,100 km²) fall into this category, but a government estimate of the forest and woodland (defined as tree cover greater than 10%) found in this category is actually only 6 km², or about 0.2% of the designated forest land area (Government of Lesotho, undated).

The areas of Escarpment Woodland are relicts of the forests which once covered these mountains and can still be seen to the north in Natal. Despite serious degradation due to livestock browsing and firewood collection, these small pockets of bushy woodland are still found in sheltered areas, valleys and dongas on the western side of the Front and Thaba-Putsoa Ranges, between 1,500 and 1,830–2,150 masl. The flora is transitional between the Highveld to the west and Drakensberg to the east. The trees are dominated by evergreens such as *Diospyros whyteana, Euclea* spp., *Halleria lucida, Ilex mitis, Maytenus* spp., *Olinia emarginata* and *Podocarpus latifolius*; and a few deciduous and semi-deciduous species such as *Leucosidea sericea, Kiggelaria africana* and *Maytenus acuminata*.

The main woody species in the Riparian Woodland are *Celtis africana, Diospyros lycioides, Rhus lancea, Ziziphus mucronata, Acacia karroo, Populus* spp. and *Salix* spp. In well-watered areas, the latter tree can reach heights of 7 m but, generally, Riparian Woodland is much lower than this.

These patches of bushy woodland represent the largest woody-biomass growing stocks in Lesotho and, as much of the woodland is evergreen, annual productivity levels are also higher than elsewhere in the country. Nevertheless areas of Escarpment Woodland are often not readily accessible, and both Escarpment and Riparian

Woodland are usually publicly or privately reserved. Both of these factors make fuelwood collection difficult, despite which, these trees are an important fuelwood resource (as has been shown by Steele and Ncholu, 1983). In a sample of 42 households, 31% collected wood from communal forests and dongas, and 14% from woodlots.

**Escarpment Grassland with Scrub Woodland**

Escarpment Grassland with Scrub Woodland is mainly found along the Western Escarpment of Lesotho, that is the Front and Thaba-Putsoa Ranges, overlooking the western Lesotho Highveld. The annual rainfall is between 800–1,000mm. The dominant vegetation is grassland with small areas of low, scrubby bush and there are also pockets of the type of bushy woodland discussed in the previous biomass class.

The land covered by this class is mainly cultivated and is situated in Berea, Butha-Buthe, Leribe, Maseru and Mohale's Hoek Districts. There are also two other areas of similar vegetation on the west-facing escarpments overlooking the Sengu River in Mokhotlong and Quthing Districts. It occurs in all districts and covers a total of 8,526 km², or 28.1% of the country.

These areas are dominated by cool-temperate grassland of varying floristic composition, depending on local environmental factors. In some areas there are scattered pockets of low, bushy *Protea* spp. Much of the Escarpment Grassland has evolved naturally although there has been significant overgrazing and cultivation.

Scrub Woodland, often known as *shallahalla*, is a product of overgrazed, degraded pasture. With continued overgrazing, soil productivity declines through soil erosion and bush encroachment ensues. The main shrub species is *Chrysocoma tenuifolia* but other important species include *Nestlera acerosa* and *Pentzia cooperi*. Together, these shrubs form an important fuelwood source known as *patsi*. Gay and Khoboko (1982) estimated that 81.8% of the sampled households in Mokhotlung use *patsi* as fuel, each utilizing an average of 23 KJ per week in the autumn. They are also an important source of summer fuel. The role of these shrubs as a fuelwood source seems to be most important in the rural villages of the Foothills Zone, and less important in the Lowland Zone (as defined by Gay 1984).

**Highveld and Riparian Grassland**

Highveld Grassland is found in western Lesotho where it is the climax vegetation. It is mainly situated in Berea, Leribe, Maseru and Mafeteng Districts, but also occurs extensively in the Orange Free State. Woody-vegetation growth is restricted by dry, frosty winters (Acocks 1975), fire (Rattray, 1960), and prolonged droughts (Ministry of Agriculture, 1979). The only woody plants to be found are thickets of stunted trees along river courses; these have been classified in the Escarpment and Riparian Woodland class. Ecologically similar areas of grassland and riparian bushes are found along the Sengu Valley in Mohale's Hoek, Mokhotlong, Qacha's Nek, Quthing, and Thaba-Teseka Districts. It is the third largest biomass class in Lesotho, covering 19.8% of the country, or 5,995 km².

Highveld and Riparian Grassland is floristically rich and forms a 25% cover grass sward about 25 to 75 cm in height. *Themeda* grass is naturally dominant but has been replaced by many other types of grasses due to overgrazing and cultivation. According to Acock's (1975) survey of South African veld types, it corresponds to the "Highland Sourveld/*Cymbopogon-Themeda* Veld Transition" and is called "Moist Cold-Temperate Grassland" by Werger and Coetzee (1983).

Riparian Woodland is commonly found along along the streams and rivers. Generally, the trees are stunted because of the severity of the climate and, more importantly, long-term over-exploitation. In Lesotho the main river valleys in this biomass class are the Mohokare, which forms the South African border to the west, and the Sengu. The main woody species, all of which take on a bush, rather than a tree, form are *Acacia karroo, Celtis africana, Diospyros lycioides, Rhus lancea* and *Ziziphus mucronata.*

In dry, low-lying and undisturbed or protected areas low bushland and scrubland no higher than 5 m can be found. This is dominated by *A. karroo, Buddleja saligna, Celtis africana, Cussonia* spp., *Diospyros* spp., *Ehretia rigida, Euclea crispa, Grewia occidentalis, Heteromorpha arborescens, Olea africana, Osyris* spp., *Rhus* spp., *Tarchonanthus camphoratus* and *Ziziphus mucronata.*

Fuelwood resources in this biomass class are low (both in terms of growing stock and annual productivity) and are very patchy in their distribution. Also, fuelwood supply from these areas of woodland and bushland is restricted by tenure patterns. Consequently the potential for fuelwood supply is low in this biomass class, which covers an area that includes important towns such as Hlotse,

Mafeteng, Quthing and Teyateyaneng, as well as the capital, Maseru.

**Alpine/Sub-Alpine Grassland and Heathland**

From the analysis of the remotely-sensed data, an Alpine and Sub-Alpine division could not be made. The combination of these two types of grassland and heathland is therefore indicative of their very low growing stocks and similar levels of productivity.

Although this situation exists throughout the Alpine and Sub-Alpine belts in Lesotho, there are other areas of the country that have been included in this biomass class. These are found on the Thaba-Putsoa Range and in the Highveld and are areas within the grassland communities that suffer from very low productivity, the most likely explanation being severe erosion. True Alpine/Sub-Alpine Grassland and Heathland are mainly found in Mokhotlong, Quacha's Nek and Thaba-Tseka Districts. Most of the areas of degraded, low-productivity grassland are in Berea and Maseru Districts. The total area covered by this biomass class, which is the largest in Lesotho, is 14,463 km², that is a little under half of the entire country (47.6%).

The sub-alpine belt in Lesotho is found between 1,830 and 2,865 masl. The soils are very thin and stony, and the climate is severe. Rainfall is variable (between about 500 and 1,200 mm) and much of it falls as snow. Sub-zero temperatures are common in winter, but in January the mean temperature is usually around 15°C. There is a frost-free period of about 185 days.

Fire-controlled *Themeda-Festuca* grassland dominates the sub-alpine belt but its species differ with aspect and altitude. Another characteristic is the occurrence of grassland dominated by *Chryscocoma tenuifolia* which covers about 13% of this zone. This is important as it is thought to be indicative of overgrazing. There is very little woody vegetation in the sub-alpine belt and, consequently, there is very little fuelwood potential.

Alpine vegetation occurs in more severe conditions than the sub-alpine grasslands and is dominated by homogeneous, low woody heathlands. The dominant genera, *Erica* and *Helichrysum*, are interspersed with grassland. They are all evergreen woody plants adapted to the cold, dry climate and all show a reduction in height that corresponds with an increase in altitude. There are a variety of vegetation types that are related to different environmental situations (e.g., bogs, pools, stream banks, wet and dry meadows, and cliffs) but almost all are dominated by grasses, mosses and herbs. Only the heath communities have a high proportion of woody-grass species, but even in the best of these communities (which are restricted to the summits of the Drakensberg) the woody species are rarely more than 1 m tall. Isolated patches of scrub which are often only 2 m in height and dominated by *Buddleja corrugata*, *Leucosidea sericea* and *Passerina montana*, can be found in undisturbed areas on the lower mountain slopes. There are effectively no fuelwood resources in the alpine belt.

**BIOMASS SUPPLY**

It has already been shown that the woody-biomass supply base is so severely restricted that only 34% of total energy consumption is supplied from woody sources. In fact based on growing stock and productivity, Lesotho has the lowest potential for fuelwood supply throughout the entire SADCC region. Nevertheless, it is useful to analyse the distribution of the existing woody-biomass resources. With the estimated population of 1,599,000 (1986) increasing due to the resettlement of mine-wage labourers in Lesotho, the existing woody biomass per capita is 1.47 tonnes per person and compares unfavourably with household wood demands (see Table 5.1).

This is obviously a very low per capita estimate and the situation is in fact quite varied. If the biomass class and district data are examined together, then two biomass classes – Highveld and Riparian Grassland, and Alpine/Sub-Alpine Grassland and Heathland – have an extremely small woody-biomass resource base. This

**Table 5.1**  ESTIMATES OF HOUSEHOLD CONSUMPTION OF FUELWOOD IN LESOTHO

| Study | Fuelwood (t/yr) |
|---|---|
| Best (1979) | 1.5 |
| Gay and Khoboko (1982) | 1.25 |
| Gay (1984) Peri-urban | 0.55 |
| Lowlands | 0.70 |
| Foothills | 1.70 |
| Steele and Ncholu (1983) | 1.8 to 2.7 |
| Wickstead (1984) | 1.1 to 1.5 |

creates a serious problem in woody-biomass supply. These are the two largest classes covering 67.4% of Lesotho (see Table 5.2). Escarpment and Riparian Woodland, with the highest level of growing stock (81.9%) and MAI (70.8%), is restricted to less than 5% of the country (see Table 5.3).

Furthermore, access to much of the Escarpment Woodland is restricted and both this type and Riparian Woodland are subject to strong regulation by the Forest Act II of 1978. This act provided the Lesotho government with the power to establish forest reserves on all land containing trees planted prior to 1973. The only other woody-biomass resource is the Scrub Woodland which regenerates in overgrazed areas. This covers 8,525 km² or 28.1% of the country (see Table 5.2).

**Table 5.2** LESOTHO: SUMMARY OF GROWING STOCK AND MAI DATA

| Biomass Class | Area (km²) | (%) | Growing Stock (1,000 t) | (%) | MAI (1,000 t) | (%) |
|---|---|---|---|---|---|---|
| Escarpment and Riparian Woodland | 1,371 | 4.5 | 1,924.9 | 81.8 | 20.5 | 70.8 |
| Escarpment Grassland with Scrub Woodland | 5,539 | 18.2 | 276.9 | 11.8 | 5.6 | 19.2 |
| Escarpment Grassland with Scrub Woodland (on small farms) | 2,987 | 9.9 | 149.4 | 6.4 | 3.0 | 10.0 |
| Highveld and Riparian Grassland | 5,995 | 19.8 | 0 | 0 | 0 | 0 |
| Alpine/Sub-Alpine Grassland and Heathland | 14,463 | 47.6 | 0 | 0 | 0 | 0 |
| TOTAL | 30,355 | | 2,351.2 | | 29.1 | |

Attempts to ameliorate the fuelwood shortage by afforestation and plantation programmes led to the instigation of the Woodlot Project in 1973, which anticipated covering an area of 3,500 ha by 1982. The most successful trees, *Eucalyptus* spp. and *Pinus radiata*, are estimated to have given initial yields of 16 t/ha (tonnes per hectare) in the north, 9.6 t/ha in the central area, and 4–5 t/ha in the south. These woodlots are categorized mainly in the Scrub Woodland biomass class. However, a little over a third of this class (35.04%) is found on small cultivated plots and access is severely restricted to anyone other than the farming family.

The few woody-biomass resources available are concentrated in six districts: Berea, Butha-Buthe, Leribe, Maseru, Mohale's Hoek and Mokhotlong. These comprise 61.6% of the country and hold 90% of the growing stock and 87.06% of the MAI. Mafeteng, Qacha's Nek, Quthing and Thaba Tseka Districts all have very low proportions of growing stock and MAI compared to their areal extent (see Table 5.3).

Although all districts in Lesotho suffer from low levels of woody biomass and growing stock, and a limited capacity for exploitation due to low annual productivity, the severest problems are found in the districts listed in Table 5.4.

**Table 5.3**  LESOTHO: GROWING STOCK AND MAI DISTRIBUTION BY DISTRICT

| District | Proportion of Growing Stock (%) | MAI (%) |
|---|---|---|
| Berea | 8.9 | 9.2 |
| Butha Buthe | 18.0 | 16.4 |
| Leribe | 22.6 | 21.0 |
| Mafeteng | 1.5 | 2.4 |
| Maseru | 13.7 | 15.3 |
| Mohale's Hoek | 12.9 | 11.8 |
| Mokhotlong | 13.9 | 13.2 |
| Qacha's Nek | 0.7 | 1.1 |
| Quthing | 4.1 | 4.9 |
| Thaba Tseka | 3.6 | 4.7 |

**Table 5.4** LESOTHO: DISTRICTS WITH SEVERE
WOODY BIOMASS SUPPLY PROBLEMS

| Districts | Severity Index* | |
|---|---|---|
| Quacha's Nek | 0.09 | increasing |
| Mafeteng | 0.26 | severity |
| Thaba Tseka | 0.26 | |
| Quthing | 0.42 | |

Severity Index = Proportion of total growing stock in district
——————————————————————————
Proportion of total area of country

# 6. Malawi

Malawi is characterized by a marked north-south division of biomass resources. There are high potential biomass reserves north of 12° 30'S, an area mainly restricted to the Northern Region. The Southern and Central Regions, and the southern parts of the Northern Region are characterized by a much lower overall level of potential although there are limited occurrences of high biomass potential.

This bipartite distribution of biomass resources is essentially a result of human activity, in particular land clearance for agriculture and fuelwood. The climax vegetation over much of Malawi is evergreen and semi-deciduous forest, and woodland. Only on thin stony soils, old sand dunes, beach deposits around the lakes, and in the seasonally flooded depressions is the natural vegetation dry open woodland or grassland.

Land clearance for subsistence agriculture, cash-cropping and fuelwood has meant that in the more densely populated parts of the Central and Southern Regions, forests and woodland have been replaced by dry, open wooded savannahs and bushy thickets. Only on the higher mountains, and in the north of the country, are evergreen and semi-deciduous forests and woodlands still found. A consequence of this distribution of woodland is that the problems of fuelwood supply are critical in the Central and Southern Regions, but much less urgent in the Northern Region. This broadly agrees with Scobey's (1984) regional fuelwood assessment.

The World Bank (1980) estimated that biomass fuelwood supply fell short of demand in 14 of Malawi's 24 Districts. Supply statistics generated from the biomass classes in this study, but not compared to the equivalent demand statistics, suggest that fuelwood supply is low in the following 13 Districts:

i) in the Northern Region: Mzimba and Nkhata
ii) in the Central Region: Dowa, Kasungu, Lilongwe, Ntcheu and Salima
iii) in the Southern Region: Blantyre, Chichwawa, Machinga, Mangochi, Nsanje and Zomba

Seven biomass classes have been identified in Malawi (see map on p.86) using NOAA–7 GAC data. Various ecological and environmental research has been used in interpreting this imagery (Agnew & Stobbs, 1972; Condliffe, 1978; Hursh, 1960; Ingram, 1984; Jackson, 1969; Shaxson, 1977) as well as field work in 1986. The recently published "National Atlas of Malawi" (Malawi Government, 1985) is an excellent, up-to-date source of thematic maps with commentaries. The biomass classes are:

- Evergreen/Semi-Deciduous Forest and Woodland
- Seasonal Open-Canopy Miombo Woodland
- Miombo Woodland with extensive Tobacco Cultivation
- Dry, Open-Canopy Miombo Woodland and Cultivation Savannah
- Mopane Woodland
- Plantations
- Swamp Woodland and Grassland

**DESCRIPTIONS OF INDIVIDUAL BIOMASS CLASSES**

**Evergreen/Semi-Deciduous Forest and Woodland**

This biomass class includes all of the evergreen forests and woodlands in Malawi and also some of the grasslands which exhibit little seasonal moisture-stress. Evergreen/Semi-Deciduous Forests and Woodlands are found extensively to the north of the Nyika Plateau. To the South they are restricted to the following areas of higher ground:

- the Viphya Mountains
- the Rift Valley Escarpment between Dowa and the Viphya Mountains
- the Mchinji Mountains
- the Dzalanyama Range
- the area between Dedza aand Mtakataka
- along the Mozambique border south of Ntcheu

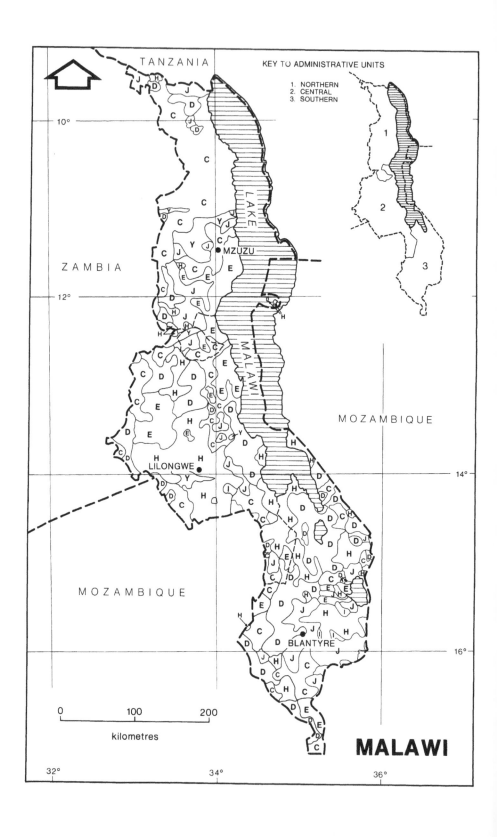

TANZANIA

KEY TO ADMINISTRATIVE UNITS

1. NORTHERN
2. CENTRAL
3. SOUTHERN

ZAMBIA

LAKE MALAWI

MZUZU

MOZAMBIQUE

LILONGWE

MOZAMBIQUE

BLANTYRE

0      100      200

kilometres

MALAWI

- the Mangochi Hills
- the Zomba Plateau
- the Mulanje Massif

A further area of evergreen grassland, the Elephant Marsh along the southern Shire Valley, falls into this biomass class. This class is found in all three regions and covers about 35,200 km² or 30% of the country.

There is a distinct seasonality in production but the overall levels are generally high. During the wet season, production levels are at their greatest and from January until June, NDVI always exceeds 175, at times reaching about 190. There is a slight decline in the NDVI values between July and October, when they range from 160 to 170, indicating slightly lower levels of productivity during the dry season.

On the high plateau and in the mountainous areas of Malawi wet evergreen forest and woodland undoubtedly constitutes the climax vegetation. These evergreen forests and woodlands have been extensively destroyed in order to provide timber and fuelwood, and to obtain land for cultivation. This is particularly so in central and southern Malawi. Climax montane forest is now only found in isolated areas and the only extensive areas of wet woodland are found in the Northern Region. Many of these forest and woodland remnants are now reserved, and all of the large Malawian forest reserves, except Philongwe, are included in this biomass class. The total Reserved Forest and Plantation area is 38,205 km² (see Tables 6.1 and 6.2).

If the main area of softwood plantations (which are identified as a separate biomass class) is subtracted from the total Reserved Forest and Plantations, and the remainder is split between the various woodland types on the basis of their geographical distribution, it is estimated that 64.6% of the biomass class is under reserved forest status. As this is ecologically the most productive wood source in Malawi, the implications of restricting access to these resources are far-reaching.

Areas of seasonal rain forest and montane grassland are found in the mountains. Montane grassland is more extensive than the forest communities and dominates the Mulanje Massif, the Zomba and Nikya Plateaux, and the Viphya Mountains. These two vegetation types are related and it has been suggested by many workers (e.g., Brown and Young, 1964; Chapman and White, 1964;

**Key to facing page:**

C Evergreen/Semi-Deciduous Forest and Woodland
D Seasonal Open-Canopy Miombo Woodland
E Miombo Woodland with Tobacco Cultivation
H Dry, Open-Canopy Miombo Woodland and Cultivation Savannah
I Mopane Woodland
J Swamp and Lake Vegetation/Tea and Coffee Cultivation
Y Plantations (*Eucalyptus/Gmelina*)

Jackson, 1954; and Shaxson, 1977) that the forest has degraded into grassland through repeated annual burning. Seasonal Rain Forest is now restricted to the highest mountains and only occurs in undisturbed locations, such as gully heads, and on deep soils. Seasonal droughts occur in many parts of Malawi but in the mountains these are very short. The lack of soil moisture-stress, combined with the high water-holding capacities of deep soils found in many upland plateau areas, mean that it is very difficult to distinguish between montane forest and grassland in terms of their phenology as they are both essentially evergreen communities.

Both natural communities and plantations are found in the forested areas. Montane rain forest is now rare in Malawi and occurs only on the wetter eastern slopes of the higher mountains between 1,200 and 2,500 masl. Mean annual rainfall usually exceeds 1,250 mm and any dry-season effects are countered by mists which supress moisture deficits. The forest is well-structured,

**Table 6.1** MALAWI: THE FOREST RESOURCE: AREAS AND INCREMENT

|  |  | Area (ha) | Increment (m³ solid/yr) |
|---|---|---|---|
| 1. | Forest Reserves (excluding areas under plantation): 75% stocked | 885,955 | 797,360 |
| 2. | Parks and Game Reserves: 75% stocked | 1,070,000 | 898,800 |
| 3. | Woodland on Customary Land: 70% stocked | 1,760,000 | 1,478,400 |
| 4. | Plantations: Public ownership | | |
|  | within Forest Reserves: softwoods | 70,686 | 1,245,044 |
|  | hardwoods | 13,359 | 184,590 |
|  | outside Forest Reserves: hardwoods | 3,934 | 36,340 |
| 5. | Plantations: Private ownership | | |
|  | outside Forest Reserves: hardwoods | 15,055 | 212,835 |
| 6. | Research Plots: mixed species | 1,500 | 18,000 |
|  | TOTAL | 3,820,489 | 4,871,369 |
|  | Total Exploitable Resources (excluding 2 and 6) | 2,748,989 | 3,954,569 |

Source: International Forest Science Consultancy (Edinburgh), unpublished.

**Table 6.2** MALAWI: PLANTATION AREAS, MAI AND INCREMENT

| | Area (ha) | MAI (m³/ha/yr) | Increment (m³solid/yr) |
|---|---|---|---|
| **1. WITHIN FOREST RESERVES** | | | |
| *Public Ownership* | | | |
| Conifers: Timber Plantation Project | 19,128 | 14 | 267,792 |
| Conifers: Viphya Plateau | 51,088 | 19 | 970,672 |
| Conifers: Nyika National Park | 470 | 14 | 6,580 |
| *Euc. grandis*: Timber Plantation Project | 1,350 | 25 | 33,750 |
| *Euc. grandis*: Viphya Plateau | 2,050 | 25 | 51,250 |
| *Euc. camaldulensis* and *Euc. tereticornis*: Rural | | | |
| Fuelwood and Pole Project | 800 | 10 | 81,000 |
| *Euc. camaldulensis*: Wood Energy Project | 9,159 | 10 | 91,590 |
| | 84,045 | – | 1,503,264 |
| | | | |
| **2. OUTSIDE FOREST RESERVES** | | | |
| *Public Ownership* | | | |
| *E. camaldulensis* and *E. tereticornis*: | | | |
| Local Authority Plantations | 2,434 | 10 | 24,340 |
| *Gmelina*, Lilongwe City Council | 1,500 | 8 | 12,000 |
| | 3,934 | – | 36,340 |
| | | | |
| *Private Ownership* | | | |
| *E. grandis*: Chombe Tea Estate | 250 | 25 | 6,250 |
| *E. grandis*: ITG | 2,225 | 25 | 55,625 |
| Euc. species: KFCTA | 4,750 | 12 | 57,000 |
| Euc. species: Press Farming | 2,830 | 12 | 33,960 |
| Euc. species: Other Estates (Estimate) | 5,000 | 12 | 60,000 |
| | 15,055 | – | 212,835 |
| | | | |
| **3. RESEARCH PLOTS** | | | |
| Mixed species | 1,500 | 12 | 18,000 |
| TOTAL | 104,534 | – | 1,770,439 |

Source: International Forest Science Consultancy (Edinburgh), unpublished.

with canopy species reacing a height of between 24 and 45 m. A middle tree stratum is found between 14 and 30 m. Both the upper and middle stratum are open but the lowest tree stratum, of between 6 and 15 m tall, forms a dense closed canopy. The herb and grass layer is poorly developed. The main tree species are *Aningeria adolfi-friedericii, Chrysophyllum gorungosanum, Cola greenwayi, Cylicomorpha parviflora, Diospyros abyssinica, Drypetes gerrardii, Entrandophragma exscelsum, Ficalhoa laurifolia, Mitragyna rubrostipulata, Myrianthus holstii, Ochna holstii, Ocotea usambarensis, Olea capensis, Parinari excelsa, Podocarpus latifolius, Prunus africanus, Strombosia scheffleri, Syzygium guineense* subsp. *afromontanun* and *Tabernaemontana johnstonii.*

In areas at higher altitudes, or with lower rainfall, or increased disturbance, the montane forests are shorter in height and are dominated by *Apodytes dimidiata, Halleria lucida, Ilex mitis, Nuxia congesta, Ocotea bullata, O. kenyensis, Podocarpus falcatus, P. gracilior, Prunus africanus* and *Rapanea melanophloeos.*

Apart from the widely distributed general types of Montane Forest in Malawi, there are forest stands dominated by a single species. On the Niyka Plateau are forests of pure stands of *Juniperus procera* and *Hagenia abyssinica.* The *Juniperus* forest is formed on slightly drier slopes than the rain forest, between 1,800 and 2,900 masl, and reaches 30 m in height. There is evidence to suggest that it is controlled by fire. The *Hagenia* forest has similar altitudinal and precipitation controls to the *Juniperus* Forest and forms a canopy 8–15 m tall on the forest margin but is not fire tolerant. On the Mulanje Massif, between 1,525 and 2,135 masl, forests of Mulanje Cedar (*Widdringtonia cupressoides*) are found, 25 to 40 m high.

Montane grassland is common throughout the Malawian highlands and is areally more extensive than all montane-forest types combined. It is dominated by grasses and small, fire-resistant *Protea* spp. shrubs. This bushy component increases in undisturbed grasslands. On the northern Viphya Plateau, Dedza Mountain and the Mulanje Massif, *Arundinaria alpina* bamboo thickets associated with bushy trees similar to those found in the montane forests, are commonly encountered.

The forest reserves in Malawi are mainly reservations of existing areas of montane forest or closed-canopy woodland on the lowlands. Their stocking rates are 70–75% and they account for 37,150 km². There are also locally important exotic timber plantations although they only account for 1,030 km² (see Tables 6.1 and 6.2). Pine plantations, mainly of *P. kesiya, P. patula* and *P. oocarpa,* have been planted for softwood timber and pulpwood. Some of these softwood plantations are classified in this biomass class (e.g., those on Zomba Mountain) but the most important coniferous plantations in the southern Viphya Mountains have been identified as a separate biomass class. Eucalyptus plantations, mainly of *E. camaldulensis, E. grandis* and *E. tereticornis* with some *Gmelina arborea*, have been planted for poles and fuelwood for curing tobacco (Forestry Research Institute of Malawi, 1985).

Semi-evergreen forests are found on the Kandoli Mountains, Thyolo Mountain, the Mulanje Massif and Zomba Plateau. These are closely associated with both the montane forests on the hills and the miombo woodlands found on the uplands. They are an intermediate vegetation type and are dominated by *Brachystegia spiciformis* but, unlike the miombo woodlands on the lowlands and plateaux, they have a thick, evergreen shrubby under-storey and the grasses are poorly developed.

In much of northern Malawi, on the Viphya and Thyolo Mountains, and in the Shire Highlands dense, closed-canopy miombo woodlands with high levels of biomass and low seasonality are found. These miombo woodland types take on two forms: first, closed-canopy woodland in areas with precipitation exceeding 1,300 mm; and secondly, a more open canopy woodland in the regions with medium to high rainfall. The canopy can reach 15 m and in some cases two canopies are present, possibly indicating previous woodland clearance. The closed-canopy woodlands are dominated by *Brachystegia longifolia* and *B. spiciformis*, but in the open woodlands *B. longifolia* is less important and *B. spiciformis, B. boehmii, B. manga, B. stipulata, Julbernadia globiflora, J. paniculata* and *Isoberlinia tomentosa* dominate the canopy. The under-storey is of very mixed composition.

Edwards (1981) analysed mature and relatively undisturbed miombo-woodland stands in the Chikala Hills near Malosa. The main results, typical of mature Miombo woodlands in Malawi, are summarized in Table 6.3.

The wet grassland of the Elephant Marsh is also classified in this biomass class, mainly because of the lack of seasonal variation in water levels, and its evergreen nature. The area is characterized by low woody biomass, except for *Borassus* and *Hyphaene* palms, and is dominated by grasses such as *Typha australis, Vossia cuspidata, Pennisetum purpureum, Cyperus papyrus* and *Echinocloa pyramidalis*.

Although production is high in this biomass class it needs to be

**Table 6.3** SUMMARY OF MATURE MIOMBO
WOODLAND NEAR MALOSA,
CENTRAL REGION

| | | |
|---|---|---|
| *Canopy* | Layers: 2 | |
| | Main constituents: | *Acacia goetzei* |
| | | *Albizia antunesiana* |
| | | *\*Brachystegia boehmii* |
| | | *\*B. bussei* |
| | | *B. spiciformis* |
| | | Other *Brachystegia* spp. (3) |
| | | *Burkea africana* |
| | | *\*Julbernardia globiflora* |
| | | *\*Pterocarpus angolensis* |
| | Trees/ha: 240 | |
| | Basal area (stems > 5 cm dbh) | |
| | m²/ha: 9.65 | |
| *Under-storey* | Layers: 1 | |
| | Main constituents: | *Bridelia cathartica* |
| | | *Combretum fragrans* |
| | | *Dalbergia nitidula* |
| | | *\*Dalbergiella nyasae* |
| | | *Diospyros kirkii* |
| | | *\*Diplorhynchus condylocarpon* |
| | | *Lannae schimperi* |
| | | *Mundulea sericea* |
| | | *\*Pseudolachnostylis maprouneifolia* |
| | | *Strychnos innocua* |
| | | *\*Terminalia stenostachya* |
| | | *\*Xeromphis obovata* |
| | | *\*Ximenia caffra* |
| | Trees/ha: 250 | |
| | Basal area (stems >5 cm dbh) | |
| | m²/ha: 2.15 | |

from Edwards (1981)
*Trees occurring in > 80% of sample plots.

ery carefully assessed in terms of fuelwood. First, the woody growing stock is very variable. In the forests and woodlands it is high, but in the areas of montane and swamp grassland it is very low. Secondly, in the forest much of the useful fuelwood is restricted to the high tree-crowns and this can pose problems of physical access. In addition to these natural factors there is the problem of access to reserved woodlands and plantations. Overall, some parts of the biomass class, particularly miombo woodland in the north, have large and relatively accessible fuelwood resources but much of the area, despite its high levels of production and growing stock, is inaccessible.

**Seasonal Open-Canopy Miombo Woodland**

Seasonal Open-Canopy Miombo Woodland is found scattered throughout Malawi and is locally dominant in some areas. It mainly occurs adjacent to Lake Malawi between Chinthetche and Nkthotakota; in the Bua and Dwangwa Valleys; at the foot of the Rift Valley Escarpment; in the upper and middle Shire Valley; and in the Shire Highlands. Most of it is contained within Central and Southern Regions and covers 24,697 km², a little over 21% of the country.

To the north, this biomass class is dominated by a relatively dense, open-canopy woodland which exhibits strong seasonality, resulting in a semi-evergreen nature. Production is high from November until April, when the NDVI is about 175–180. After April there is a dramatic drop in production and in August the NDVI values are only about 150, indicating very low levels of productivity in August and September. The main difference between the Seasonal Open-Canopy Miombo Woodland and the Evergreen Miombo Woodland in the north is the much lower level of biomass productivity during the dry season. This seasonality can be caused by either climatic or pedological factors.

In the south, the woodlands are far more open in character. Whilst some of them can be termed open-canopy woodland, they commonly grade into thickets and wooded savannahs. This poorer vegetation in the south is due to the extensive clearance of the original Seasonal Open-Canopy Miombo Woodland for cultivation and fuelwood.

Undisturbed Seasonal Open-Canopy Miombo Woodland, such as that found in northern and central Malawi, has an open canopy dominated by *Brachystegia boehmii*, *B. manga*, *B. stipulata*, *B. spiciformis*,

*Julbernadia globifolia, J. paniculata* and *Isoberlinia tomentosa*. Once cleared, the nature and composition of the woodland changes drastically and in the south of the country, three different types of vegetation can be seen.

Slightly better wooded areas occur on soils which have been cultivated but are still relatively fertile. Here, patches of relict miombo woodland and isolated trees are found. These are mainly *Adansonia digitata, Acacia albida, Cordyla africana* and *Stericulia* spp. There are also palms between the grasslands, particularly *Hyphaene ventricosa*. In less fertile areas, the vegetation degrades into either a wooded or thicket savannah. Thicket savannah occurs on alluvial soils in the lower Bwanje and Shire Valleys, and on the southern shores of Lake Malawi. The tree species are similar to those found in wooded areas but all grow in a much more stunted form. On thin or stony soils, a very open wooded savannah is found. This is dominated by *Brachysytegia* spp., *Combretum* spp., *Pterocarpus* spp. and *Colophospermum mopane*.

Seasonal Open-Canopy Miombo Woodland has a very variable fuelwood potential. In the north it is characterized by high levels of growing stock and productivity, particularly in the wet season, and it is an important fuelwood reserve. The degraded forms in the south have lower fuelwood potential but are none the less locally important.

**Miombo Woodland and Extensive Tobacco Cultivation**

Well-wooded, open-canopy miombo woodland is often found where there is extensive tobacco cultivation between areas of woodland. It can be distinguished as a separate biomass class on the basis of the combined phenological characteristics of the woodland and the tobacco crop. It occurs on the Zambian border to the south of the Niyka Plateau, on the Kasungu Plateau, the southern parts of the Dedza-Ntcheu Highlands and in the southern part of the Zomba Plateau. In total it covers about 16,624 km² or about 14% of the total land area.

This class is floristically and structurally similar to the other types of miombo woodland, and field observations suggest it is usually an intermediate type between the Seasonal Open-Canopy Miombo Woodland and the Dry Open-Canopy Miombo Woodland. However, when examining its phenological characteristics which are one of the main criteria for its differentiation, it is important to include the growth cycle of tobacco. This is because in the areas covered by this biomass class there is an intricate intermixing of woodland and tobacco cultivation which cannot be differentiated with 8 km resolution data.

The phenological characteristics and overall productivity of the woodland component is typical of the seasonal woodlands of Malawi. The overall level of productivity of tobacco cultivation is lower than that of the woodlands, although quite high for a ground crop.

Wet-season productivity levels are slightly lower than those of Seasonal Open-Canopy Miombo Woodland; NDVI values reach only 170 to 175. Nevertheless, production is still relatively high. Again, there is a decline in dry-season production levels, but due to the beneficial effects of the green tobacco crop this is not as drastic as in the Seasonal, Open-Canopy Miombo Woodland areas. However there is a long fall in productivity, lasting from July to October. There is also a far more marked seasonality here than in the Seasonal Open-Canopy Miombo Woodland, which is due to the areal proportion of the tobacco fields in the class. The seasonal variations in productivity are similar to those of Dry Open-Canopy Miombo Woodland.

The woodland canopy is dominated by the same species as the Seasonal Open-Canopy Miombo Woodland, namely *Brachystegia boehmii, B. manga, B. stipulata, B. spiciformis, Julbernadia globifolia, J. paniculata* and *Isoberlinia tomentosa*. However as it is slightly drier and has a lower biomass, the canopy is less dense and lower than that of the Seasonal Open-Canopy Miombo Woodland. *B. boehmii* is the dominant tree species in most areas and tobacco is grown as a monocultural row-crop with no woody component.

In terms of fuelwood production this class is similar to Seasonal, Open-Canopy Miombo Woodland. However, the annual productivity levels may be slightly lower. It is, of course, an important and accessible source of fuelwood for tobacco curing and, in many areas, the woodlands are owned by the tobacco estates which leads to the problem of restricted access.

**Dry Open-Canopy Miombo Woodland and Cultivation Savannah**

Dry Open-Canopy Miombo Woodland and Cultivation Savannah is found extensively in the Southern and Central Regions. Large areas are found on the Lilongwe Plain, in the Banjwe Valley, on the Shire Highlands and on the Phalombe Plains. Smaller, but nevertheless important, areas occur in the Dwangwa Valley, in the lower

Shire Valley, on the lakeshore near Salima and to the north of Lake Chilwa. In total this biomass class covers 20,252 km², about 17.2% of Malawi.

It suffers a severe dry-season hiatus in production, lasting from July until October, during which NDVI values never exceed 150. Wet-season production is relatively high, however, and the NDVI values between January and May are relatively constant at around 175. There are sharp rises and falls between the two periods of high and low levels of production.

Most of these areas are characterized by a very open woodland or a "cultivation savannah" (a term commonly used in Malawi to describe a savannah of maize cultivation and grassland with miombo woodland and exotic tree species). It generally grades from very open woodland to a grassy savannah with few trees. In the most severely deforested areas, mangoes are often the only trees left. This class develops mainly after woodland has been cleared for cultivation and fuelwood. However, on very thin, stony soils Dry, Open-Canopy Miombo Woodland is the natural vegetation. In the north, areas of mixed woodland thicket are found.

Very open miombo woodland develops naturally on thin stony soils with little capacity to hold water. This is found on the Rift Valley Escarpment, where it is known locally as *msuku* woodland. The main canopy dominants are *Brachystegia* spp., especially *B. boehmii*.

In most areas, however, Dry Open-Canopy Miombo Woodland is caused by the clearance of well-developed woodland. Under increased population pressure, it grades into Cultivation Savannah. In the Lilongwe Plateau (Central Region), degraded miombo woodland forms a Cultivation Savannah in which the only remnants of open-canopy woodland are sacred graveyards found amongst grassland and maize fields. The woody component of the Cultivation Savannah situated away from these remnants is dominated by mangoes (*Mangifera indica*), *Acacia* spp., *Combretum* spp., *Piliostigma* spp. and *Uapaca kirkiana*. It is particularly common on the Lilongwe Plain, and in the Namwera and Malindi regions.

In the Southern Region, population pressure in the areas of open, dry miombo woodland has been far greater than further north, and Cultivation Savannah and woody thickets are more commonly found. Fertile soils are characterized by grasslands interrupted by copses of relict woodland as well as isolated trees such as *Adansonia digitata, Acacia albida, Cordyla africana, Hyphaene ventricosa* and *Stericulia* spp. If the soils are less fertile, either a thicket or a savannah can form. Thickets occur on alluvial soils to the south of Lake Malawi and the species they contain are similar to those found in wooded areas although the trees take on a more stunted form. On thin stony soils, a wooded savannah is found which is dominated by *Brachystegia* spp., *Combretum* spp., *Pterocarpus* spp. and *Colophospermum mopane*. Edwards (1982) examined savannah woodland which had been disturbed near Blantyre. Direct comparisons can be made with Edwards's (1981) study of a mature miombo woodland. The disturbed, savannah woodland lacks stratification, has a significantly lower basal area (4.87 m²/ha compared to 11.8 m²/ha) and a better developed grass layer. The main dominants are *B. boehmii, Burkea africana, Julbernardia globiflora, Monotes africanus* and *Pseudolachnostylis maprouneifolia*.

To the east of the Vwaza Marshes and on the Nkhamanga Plains, this biomass class is represented by low woody-thicket vegetation with extensive grassland areas. The main thicket-forming tree species are *Acacia* spp., *Combretum* spp. and *Reissantia indica*.

There are two reasons why fuelwood potential in this biomass class is low. First, the restriction placed on annual production by reduced dry-season productivity; and secondly, a low level of growing stock due to previous clearance. There are also access problems as much of the biomass class occurs in intensively farmed areas.

**Mopane Woodland**    Southern Malawi is at the northernmost extent of the Mopane Woodland range and so this class only occurs in a few scattered areas of the country, concentrated in the Bwanje and Shire Valleys. Consequently, it is almost entirely found in the Southern Region and the areal extent is 864 km², only 0.7% of the country.

It is phenologically distinct from the Miombo Woodland. Levels of biomass productivity are high in January, when NDVI is about 180, followed by a long slow decline until the period of lowest production is reached in September and October with NDVI values of about 145. Overall levels of productivity are therefore generally much lower than those of miombo woodland.

The distribution of this biomass class is strongly influenced by the occurrence of compacted, sandy alkaline soils which are vulnerable to degradation. These are locally classified as *mopanosols* and are generally left uncultivated. The Mopane Woodland that develops on them forms an open-tree savannah with *Colophospermum mopane* dominating the canopy.

Fuelwood resources in the Mopane Woodland areas are lower than in the miombo woodlands because of lower annual levels of production. Nevertheless, the overall growing stock is greater than the surrounding vegetation. The combination of these two factors could lead to local over-exploitation.

**Plantations**

Plantation areas in the Northern and Central Regions can be differentiated from the Evergreen/Semi-Deciduous Forest and Woodland in terms of their biomass and phenology. The most extensive occurrences of this class are found to the south of the Nyika Plateau and around Lilongwe. The total area is about 4,238 km², or 3.6% of the country.

Productivity is high and exhibits a strong seasonality. Levels peak in March when NDVI is about 190, followed by a strong decline in NDVI values through to August when they are only about 150. After this they show a rapid increase. These areas are mainly dominated by coniferous and other fast-growing softwood trees. The southern Viphya Plateau is an area of pine plantations, mainly *P. kesiya, P. patula* and *P. oocarpa*. This is essentially a pine monoculture. The other large area is found around Lilongwe and is dominated by the many *Gmelina arborea* plantations established in the early 1970s by the Capital City Development Corporation (CCDC). Most of these are now mature but many have not yet been harvested.

Levels of productivity in this biomass class are high and it represents a vast untapped store of biomass reserves. However, apart from some peripheral (and usually illegal) exploitation, the area is almost entirely reserved.

**Seasonal Swamp Woodland and Grassland/Tea and Coffee Plantations**

Swamp Woodland and Grassland are developed most extensively around Lake Chilwa on the Mozambique border. This type of vegetation is, however, common throughout Malawi in the seasonally flooded depressions known locally as *dambos*. Although individual *dambos* are too small to be mapped from AVHRR 8 km data, it should be borne in mind that this type of vegetation is locally important throughout the country.

In the Southern Region this biomass class also includes the extensive but ecologically unrelated areas of tea and coffee plantations around Thyolo and Mulanje. The ecologically strange combination of swamp vegetation with tea and coffee plantations is due to their similar levels of productivity and phenology. Both have relatively high levels of wet-season production which decline only slightly in the dry season. This high productivity and negligible seasonality is due to the large water-holding capacity of the swamp soils. In the plantation areas, this is due to intensive tea and coffee production, and the extensive evergreen coniferous woodlots (the wood being reserved for tea processing). Consequently, this biomass class is found in all regions; it covers an area of 15,539 km² or 13.2% of the country.

Fringing Swamp Woodland is found on a variety of soil types characterized by seasonal flooding. Overall levels of production are significantly lower than those of the surrounding woodlands. Wet-season productivity is relatively high, with NDVI values of about 165 occurring between March and May. This is followed by a lower dry-season productivity in September and October, with NDVI values of about 145. On sandy soils the woodland is dominated by *Terminalia sericea* which grades into *Brachystegia boehmii*-dominated miombo woodland. As the sandy soils have low water-holding capacities, the woodland is marked by a strong seasonality. On the finer textured soils, found along the larger river courses, *Terminalia sericea* is replaced by *Acacia* spp. and *Combretum* spp. Here, vegetation seasonality is much less marked. The areas of seasonal grassland are dominated by a tall and dense grass cover with very few trees, the dominant species being *Acacia seyal* in wet areas, and *A. spirocarpa* in dry areas.

In many places, the seasonal vegetation communities around swamps and *dambos* have been extensively modified by farming. The grasslands are kept low by cattle grazing and the wetter soils are used for dry-season cultivation. Fuelwood resources are low due to the low levels of wood-growing stock, despite high rates of productivity. Localized pressure on fuelwood resources is seen in the rapid destruction of the few trees in these ecosystems.

The plantation areas are dominated by monoculture crops. Tea and coffee are dominant but other tropical shrubs used for nut and oil production are found in the area to the south of Blantyre and Limbe. As most of the plantations are of shrubs rather than ground crops, they exhibit a phenology which is more related to trees than ground crops and grasslands. There are a few areas of woodland, especially in the tea-growing areas around Thyolo. These are mainly coniferous plantations, the wood being used on the estates.

Therefore, fuelwood resources in these regions are low with the only extensive woodland being coniferous plantations on tea estates. Productivity levels are quite high; the main problem here is access to the available wood resources.

**BIOMASS SUPPLY**

Malawi has substantial biomass reserves of 367.8 million tonnes of growing stock, which equals 50 tonnes per person (based on 1986 population estimates). The mean annual increment is also quite high. Nevertheless, Malawi experiences severe biomass supply problems in a number of areas, which can be attributed to four main causes.

First, biomass resources are poorly distributed with respect to population. Almost 50% of the biomass reserves occur in the Northern Region, 30% in the Central Region, and only about 20% in the Southern Region (see Table 6.4). The population is concentrated

**Table 6.4**  MALAWI: GROWING STOCK AND MAI DISTRIBUTION BY DISTRICT

|  | Proportion in each region | |
|  | Growing Stock (%) | MAI (%) |
| --- | --- | --- |
| Central | 34.3 | 33.4 |
| Northern | 44.3 | 47.9 |
| Southern | 21.4 | 18.7 |

in the latter two regions, and 52% of Malawi's total fuelwood demand is generated by domestic consumption and the tobacco farms of Central District. Only 7% of the demand is generated in the Northern District, which has 44.3% of the total reserves.

Secondly, the vast bulk of the biomass resources are concentrated in a few biomass classes (see Table 6.5). Despite covering 30% of the land area, the Evergreen/Semi-Deciduous Forest and Woodland

**Table 6.5**  MALAWI: SUMMARY OF GROWING STOCK AND MAI

| Biomass Class | Area (km²) | (%) | Growing Stock (Mill t) | (%) | MAI (Mill t) | (%) |
|---|---|---|---|---|---|---|
| Evergreen/Semi-Deciduous Forest and Woodland | 35,200 | 30.0 | 250.7 | 68.2 | 7.9 | 58.1 |
| Seasonal Open-Canopy Miombo Woodland | 24,697 | 21.0 | 49.0 | 13.3 | 1.2 | 8.8 |
| Miombo Woodland with extensive Tobacco Cultivation | 16,624 | 14.2 | 27.9 | 7.6 | 0.4 | 2.9 |
| Dry Open-Canopy Miombo Woodland and Cultivation Savannah | 20,252 | 17.2 | 19.2 | 5.2 | 1.0 | 3.6 |
| Mopane Woodland | 864 | 0.7 | 3.2 | 0.8 | 0.1 | 0.7 |
| Plantations | 4,238 | 3.6 | 17.9 | 4.9 | 2.2 | 26.3 |
| Swamp Woodland and Grassland | 15,539 | 13.2 | 0 | 0 | 0 | 0 |
| TOTAL | 117,414 | | 367.8 | | 13.7 | |

Class holds 68.2% of the growing stock and 57.6% of the MAI. If all of the moist woodland, forest and plantation classes are combined they account for 94.3% of the growing stock and 72.9% of the MAI, but only cover 79% of the country. Consequently, large areas exist that have very poor biomass reserves. The largest two biomass classes falling into this category are the Dry Open-Canopy Miombo Woodland and Cultivation Savannah, and Swamp Woodland and Grassland. These account for 30.4% of the land area, but only 5.2% of the growing stock and 3.6% of the MAI. The significance of the low biomass reserves in these two classes is amplified by the fact that most Malawian farmers are concentrated in these zones. The majority of the smallholder farms are located in areas of Dry Open-Canopy Miombo Woodland and Cultivation Savannah. Therefore, these are high population-density areas and, because much land is given over to maize cultivation and woodland is scarce, they represent situations of high demand and low supply. The main areas are the

- Lilongwe Plains
- Phalombe Plains
- parts of the Shire Valley

- Dedza-Ntcheu Highlands
- Southern shores of Lake Malawi and Lake Malombe

The large area of swamp vegetation around Lake Chilwa in the Southern Region is an important area of commercial swamp rice production. This is surrounded by an area of naturally lower biomass potential. Other areas falling into the latter class are the tea and coffee plantations of the Southern Region.

Another major problem associated with the woody biomass resources of Malawi is that of accessibility. Although a little over half of the biomass resources are found in Central and Southern Regions, the areas of highest potential (i.e., the woodlands and forests) are mainly reserved by the Forest Department. This accentuates the problem of biomass resources in the two most densely populated regions. The type of access restriction varies from Forest Reserves to Game Reserves and Plantations, but all restrict the collection of available fuelwood. In terms of area, the largest amount of restricted woodland and forest is that found in National Parks and Game Reserves (51.9%), followed by Forest Reserves (43%) (see Table 6.6).

**Table 6.6**  WOODLAND RESERVATION IN MALAWI

| Type of reservation | Area (Km²) | Proportion of all reserved land (%) | Number of units in each region C | N | S |
|---|---|---|---|---|---|
| 1. Parks and Game Reserves | 10,700 | 51.9 | 2 | 2 | 5 |
| 2. Forest Reserves | 8,856 | 43.0 | 17 | 20 | 27 |
| 3. Publicly-owned Plantations | 880 | 4.3 } | | | |
| 4. Private plantations | 150 | 0.7 } | 6 | 1 | 14 |
| 5. Research Plots | 15 | <0.1 | unknown | | |
| TOTAL | 20,601 | 100.0 | | | |
| 6. Accessible Woodland or Customary Land | 17,600 | 85.4* | | | |

* As a proportion of total reserved land

**Table 6.7** PREFERRED FUELWOOD AND CHARCOAL TREES

| Botanical name | Chichewa name | Qualitative Rating for: | | Main biomass classes |
|---|---|---|---|---|
| | | Fuelwood | Charcoal | |
| Acacia galpinii | Nkungu | G | – | 4 |
| Brachystegia manga | Msumbu | G | – | 1,2,3 |
| B. spiciformis | Chombe | E | – | 1,2,3 |
| B. spiculata | Chiombo | G | – | 1,2,3 |
| Bridelia micrantha | Mpasa | G | * | 12 |
| Cassine aethiopica | Mkokopa | G | – | 1 |
| Colophospermum mopane | Tsanya | G | * | 4,5 |
| Crossopteryx febrifuga | Dangwe | G | – | 3,4 |
| Erythrophleum suavelons | Mwabui | G | – | 1 |
| Gmelina arborea | Malaina | F | – | 6 |
| Parinari curatellifolia | Mbula | – | * | 2,3,4 |
| Piliostigma thonningii | Chitimbi | G | – | 4 |
| Pseudolachnastylis maprouneifolia | Msolo | – | * | |
| Pterocarpus rotundifolius | Khongozi | G | – | 3,4 |
| Schrebera alata | Mkuko | G | – | 1,2,3 |
| Securidaca tongepedunculata | Bwazi | G | – | 2,3 |
| Strychnos spinosa | M'mawaye | G | – | 3,4 |
| Sterospermum kunthianum | Kabrungiti | G | – | 4 |
| Uapaca kirkiana | Mtoto | – | * | 4 |

Adapted from Pullinger & Kitchin (1982)

Key to fuelwood rating (from Pullinger & Kitchin 1982)
E – Excellent
G – Good
F – Fair
* – Used for charcoal production

Key to biomass classes
1. Evergreen/Semi-Deciduous Forest and Woodland
2. Seasonal Open-Canopy Miombo Woodland
3. Miombo Woodland with extensive Tobacco Cultivation
4. Dry Open-Canopy Miombo Woodland and Cultivation Savannah
5. Open Woodland
6. Plantations

The quality of wood is a further factor in determining its fuelwood potential and so the quality of the main tree species used for fuelwood or charcoal needs to be taken into account. It can be seen from Table 6.7 that all the biomass classes in Malawi, with the exception of the Swamp Woodland and Grassland, have commonly occurring preferred fuelwood and charcoal tree species. This table does not indicate the frequency at which the species listed occur but commonly found species (such as *Acacia galpinii*, *Brachystegia* spp., *Gmelina arborea*, *Piliostigma thonningii*, *Pterocarpus rotundifolius* and *Uapaca kirkiana*) can be noted. A further qualification needs to be made when evaluating individual tree species for fuelwood and charcoal requirements. Most of the dominant trees mentioned above have multiple end-uses (see Table 6.8) which conflict with the use of a tree for fuel alone.

Despite Malawi's seemingly high biomass-resource potential there are significant supply problems in the following areas:

- Lilongwe Plains
- Phalombe Plains

**Table 6.8** MULTIPLE END-USES OF SOME MAJOR TREES USED FOR FUELWOOD IN MALAWI

| Tree | Main biomass classes found in: | Main end-uses |
|---|---|---|
| *Acacia galpinii* | 4 | fuelwood, timber, gum, medicine, furniture |
| *Brachystegia manga* | 1, 2, 3 | fuelwood, rope |
| *B. spiciformis* | 1, 2, 3 | fuelwood, rope, beehives, medicine |
| *B. stipulata* | 1, 2, 3 | fuelwood, rope |
| *Colophospermum mopane* | 4, 5 | fuelwood, charcoal, poles, furniture, construction timber, lime, gum, medicine |
| *Piliostigma thonningii* | 4 | fuelwood, animal/human food, soap, rope, timber, medicine |
| *Uapaca kirkiana* | 4 | charcoal, construction timber, fencing, food, drink |

Adapted from Pullinger & Kitchin (1982)

Key to biomass classes
1. Evergreen/Semi-Deciduous Forest and Woodland
2. Seasonal Open-Canopy Miombo Woodland
3. Miombo Woodland with extensive Tobacco Cultivation
4. Dry Open-Canopy Miombo Woodland and Cultivation Savannah
5. Mopane Woodland

- Dedza-Ntcheu Highlands
- Lower and Middle Shire Valley
- South Shores of Lakes Malawi and Malombe
- the coffee and tea plantations around Thyolo and Malanje
- Chizumulu and Likoma Islands

These areas of shortfall in woody-biomass resources have mainly been brought about by excess demand in areas of extensive forest clearance or with access restrictions.

The first five areas listed above all have low woody-biomass supply potential due to extensive clearance for cultivation, fuelwood and other uses. This is also true of Chizumulu and Likoma Islands in Lake Malawi. In these areas woody biomass resources are very low, too small in fact to be identified from AVHRR imagery. Wood is brought in by boat from Niassa Province in Mozambique. The coffee and tea plantations in the south are another further area with woody-biomass supply problems. Here the productive wood-lands are reserved for commercial operations, and woody-biomass demand from local farmers and plantation workers has to be met from small, ecologically marginal woodlands on the Rift Valley Escarpment to the west.

In all of the border regions – especially around Mulanje and in the Dedza-Ntcheu Highlands – the woody-biomass supply problems are reaching a critical level with the influx of refugees from Mozambique.

# 7.  Mozambique

Vegetation distribution in Mozambique is strongly influenced by certain ecological factors. These include levels of rainfall and seasonality; relief and the water-holding capacity of the soils; and botanical factors which influence the distribution of species (White, 1983). This floristic division is most noticeable between the vegetation types of southern Mozambique, and central and northern Mozambique. There is a further major division between the central and northern inland regions, and the coast. The automated classification of vegetation types has produced a series of biomass classes that have amalgamated a number of distinct vegetation types. Within each biomass class there are broad similarities in quantity, productivity and seasonality, regardless of vegetation type. This suggests that the botanical criteria previously used to classify vegetation in Mozambique have less application in fuelwood assessment than classifications based on ecological and structural criteria. In this respect, the work carried out by Malleux (1980) is very useful.

Biomass classes in Mozambique are dominated by forests and woodlands (88.5% of the country) which range from evergreen forests through semi-deciduous forests and woodlands to deciduous, savannah woodland. Altogether, they cover a far larger area than Lanly's (1981) estimate of 19.7% of the country. The wet, semi-deciduous forests and woodlands form the most extensive biomass class, accounting for 27.9% of the country.

Initially, this appears to be a healthy picture in terms of fuelwood resources. However, there are two other factors which need to be considered. First, the annual increment of wood production is very low in many of the deciduous woodlands and exploitation could easily lead to rapid vegetation degradation. A number of areas of woodland have already degraded into savannahs (with few woody plants and shrubby thicket vegetation) with low biomass reserves and productivity. This type of vegetation occurs extensively in Gaza, Inhambane, Maputo and Tete Provinces. It is also seen locally in all other provinces except Nampula, Niassa and Zambezia. Secondly, the distribution of biomass resources is very skewed. Areas of extensive high growing stock and high-productivity biomass reserves are found extensively in Nampula, Niassa and Zambezia Provinces, and over large areas of Cabo Delgado, Manica and Sofala Provinces. These areas are much more restricted in Inhambane and Tete Provinces, and rarely occur in Gaza and Maputo Provinces.

Munslow (1984) compared annual wood production with fuel needs in different provinces. Wood production was based on FAO forestry statistics (FAO, 1981) derived from wood production data from the 1970s. The more up-to-date information used in this study generally supports these data, but there are some significant modifications:

- the biomass reserves and production in Gaza and Tete are far less than previously estimated;
- the situation in Cabo Delgado is slightly worse; *and*
- production and reserves in Manica, Nampula and Zambezia are better than previous estimates suggest.

Mozambique is divided into 11 vegetation zones (see map on p.109) based on interpretation of NOAA-7 AVHRR GVI data (at 8km resolution) for the area north of the Limpopo River, and NOAA-7 AVHRR GAC data (at 4km resolution) for the area to the south. Previous ecological and forestry studies have also been taken into account. Previous work on the vegetation of Mozambique can be divided into two groups. First, that carried out by ecologists and botanists (e.g., Gomes Pedro and Grandvaux, 1956; Barreto and Soares, 1972); and secondly, foresters (FAO, 1978; Kriek, 1981).

The vegetation zones are:

Evergreen Miombo Woodland and Coastal Forest
Wet Seasonal Forest and Woodland
Seasonal Miombo Woodland
Dry Miombo Woodland
Mopane Woodland
Coastal Forest Mosaic
Dry Riparian Woodland

Degraded Agricultural Land
Lowland Sublittoral Forest and Bushland
Lubombo Hills Woodland
Littoral Grassland

## DESCRIPTIONS OF INDIVIDUAL BIOMASS CLASSES

### Evergreen Miombo Woodland and Coastal Forest

Evergreen Miombo Woodland is best developed adjacent to the coast in Zambezia Province. In many places it becomes impossible to separate it from the wetter Coastal Forest on biomass, productivity, and phenological criteria. The most extensive area of Evergreen Miombo Woodland is in Nampula and Zambezia Provinces. From southern Zambezia Province, around Quelimane, a discontinous strip of Evergreen Coastal Forest can be identified extending southwards to the Limpopo River; it is best developed in northern Sofala and southern Inhambane Provinces. The total area of Evergreen Forest and Woodland in Mozambique is about 121,002 km² or 15.1% of the country.

Evergreen Miombo Woodland suffers little seasonal moisture stress. Photosynthetic activity is high throughout the year with NDVI values varying between 165 and 185. It is highest between January and April and then declines slowly until it reaches a low in September and October. The canopy is rarely above 15 m, and usually stands at 10 m. Beneath it is a layer of low (5–10 m high) trees and a poorly developed herbaceous layer. Evergreen Miombo Woodland is floristically rich and the canopy is dominated by *Brachystegia spiciformis* and *Julbernardia globiflora*.

The coastal forest belt is mainly characterized by a dry deciduous forest on the lowland plains with moister evergreen and semi-deciduous forest on the windward mountain slopes. The evergreen and semi-deciduous component of the coastal forest complex is included in this biomass class. Much of this forest has been now converted to low woodland (Malleux, 1980) with an upper canopy of 10–12 m and a lower canopy of 5–7 m; it is dominated by *Acacia nigrescens* and *Albizia* spp. In many areas this forest has become so degraded that it now forms a thicket-like vegetation with shrubby species varying between 3 and 7 m in height, and emergents of up to 10 m tall. Not all of the low forest and thicket can be attributed to degradation as low, open coastal vegetation is the natural vegetation on sandy soils. This is particularly so in Inhambane Province.

Scattered pockets of higher, non-degraded coastal forest still

**Key to facing page:**

B Evergreen Miombo Woodland and Coastal Forest
C Wet Seasonal Forest and Woodland
D Seasonal Miombo Woodland
G Coastal Forest Mosaic
H Dry Miombo Woodland/ Wood and Shrub Thicket
I Mopane Woodland
J Dry Savannah Woodland
K Degraded Agricultural Land
N Dry Riparian Woodland
Q Lubombo Hills Woodland
U Lowland Sublittoral Forest and Bushland
V Littoral Grassland

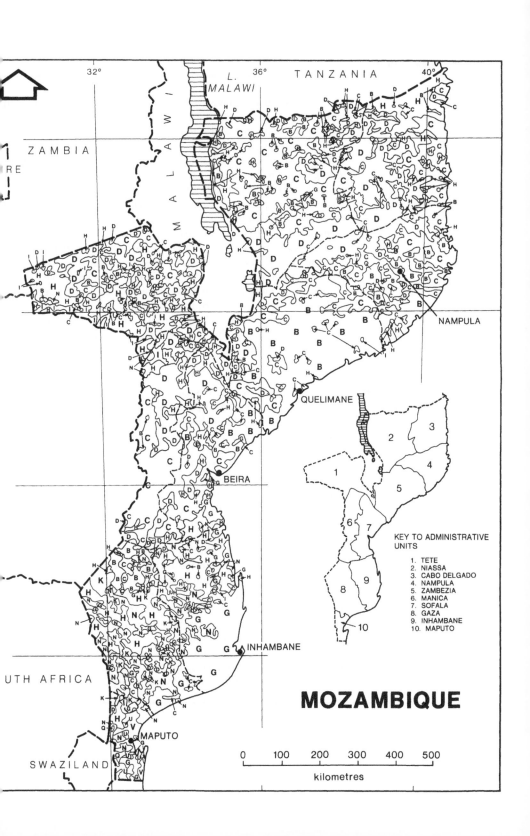

**MOZAMBIQUE**

KEY TO ADMINISTRATIVE
UNITS

1. TETE
2. NIASSA
3. CABO DELGADO
4. NAMPULA
5. ZAMBEZIA
6. MANICA
7. SOFALA
8. GAZA
9. INHAMBANE
10. MAPUTO

0    100    200    300    400    500
kilometres

exist. These forests have a canopy of about 20 m in height, with emergents reaching 40 m. The main species are *Lovoa swynnertonii*, *Maranthes goetzeniana* and *Crossonephelis (Melanodiscus) oblongus*. In Manica and Sofala Provinces, and on the windward slopes of the Macondes Plateau, the moister forest is semi-deciduous rather than evergreen. It has a canopy of about 15 to 20 m tall with emergents of up to 35 m; the mean height of the largest trees is about 25 m. The main canopy species are *Newtonia buchananii*, *Erythrophleum suaveolens* and *Pachystela brevipes*. There is also a well-developed shrub layer dominated by *Albizia adianthifolia*.

There are variations to this general pattern, particularly around the Zambezi Delta where *hirtella* forest develops as a response to high precipitation (1,200–1,400 mm) and a high water table. *Hirtella* Forest is a mosaic of miombo woodland and semi-deciduous *Pteleopsis-Erythrophleum* forest. It is dominated by *Hirtella zansibarica*.

The high growing stock and annual productivity levels mean that, in most places, the fuelwood potential of these evergreen forests is very high.

**Wet Seasonal Forest and Woodland**

This is the most extensive biomass class in Mozambique, accounting for 223,141 km² or 27.9% of the country. The range of forest and woodland types found in this biomass class (montane forest, wet coastal forest and wet miombo woodland) are all characterized by a high total biomass and a slight, but nevertheless marked, seasonality in photosynthetic activity. These forests and woodlands occur in the wetter areas (mean annual rainfall is usually greater than 1,000 mm) but are subject to drought conditions for part of the year. Photosynthetic activity is high from January until May, with NDVI values varying between 180 and 195. Activity then declines until September at which time NDVI values vary between 150 and 170. The low point in photosynthetic activity in September is followed by a sharp rise in rates during November and December.

Montane forest vegetation is particularly important in Manica, Niassa and Zambezia Provinces. Wet miombo woodland is found in Inhambane, Niassa, Sofala, Tete and Zambezia Provinces. Along the coastal zone of Cabo Delgado Province, the wetter coastal woodland is included in this class.

Above 1,200 masl the mountainous areas are dominated by montane forests. Montane vegetation is characterized by high rates of photosynthetic activity (NDVI values are generally higher than

175) and here is a slight but marked seasonality. Activity is slightly higher between January to March (NDVI values are around 180 to 190) and shows a small decline reaching its lowest values between August and October (NDVI of approximately 170 to 180). Activity then increases again towards the end of the year. The decline in photosynthetic activity, and consequently in biomass productivity, is due to the short dry season. However, this is often offset by dry-season mists.

The forests range from being well to poorly structured and are floristically diverse. They include some conifers as well as broad-leaved tree species. In high altitude areas, with a higher annual rainfall, a well-structured forest can form with a well-developed, but not dense, canopy of between 14 and 30 m. Emergents reach heights of 25–45 m (the average height being 30–38 m). A lower, dense, small tree canopy is found between 6 and 15 m. Beneath the tree strata there is a 3–6 m shrub layer and a sparse grass and fern herb layer. The main tree species found in both types of forest are *Ocotea* spp., *Podocarpus* spp. (especially *P. latifolius*), *Prunus africana* and *Xymalos monospora*. In many areas the montane forests are now much lower in height and they form tree canopies of between 8 and 15 m high. These are usually accompanied by a significant influx of secondary, deciduous species which take on a low tree and shrubby habit (Malleux, 1980).

Unlike its evergreen equivalent, wet miombo woodland experiences a short period of soil moisture-stress and its vegetation shows a noticeable deciduous element. The structure is similar to Evergreen Miombo Woodland with a tree canopy reaching about 10–15 m in height, overlaying poorly-developed shrubby and grass layers. There is little floristic difference between evergreen and wet miombo woodland; both are dominated by *Brachystegia spiciformis* and *Julbernardia globiflora*.

The lowland, deciduous coastal forests along the northern Mozambique coast also fall into this biomass class. These forests are floristically rich and have many tree species, the most important of which are *Afzelia quanzensis*, *Chlorophara excelsa* and *Sterculia appendiculata*. Other important canopy and emergents are *Albizia gummifera*, *Ekebergia capensis*, *Erythrophleum sauveolens*, *Khaya nyasica*, *Millettia stuhlmannii*, *Newtonia buchananii*, *Pachystela brevipes*, *Syzygium guineense* and *Xylopia aethiopica*. It forms a canopy at about 12.5 m with emergents reaching heights of 25 m. In addition there is a well-developed, thick closed sub-canopy consisting mainly of *Albizia adionthifolia, Bauhinia* spp., *Harungana madagascariensis, Macaranga capensis* and *Trema orientalis.*

Large proportions of the Montane Forest in this biomass class are reserved, but it has not been possible to obtain detailed current data to calculate the exact proportion. The remainder of this biomass class does provide a fuelwood source of high potential, both in terms of annual productivity and growing stock.

**Seasonal Miombo Woodland**

Seasonal Miombo Woodland is found throughout Mozambique and occurs in all provinces. The most extensive areas fringe the Macondes and Niassa Plateaux. Isolated patches are also found in areas with rainfall in excess of 800 mm, in the lower Zambezi Valley. There is a further large area on the windward flanks of the Shire Mountains. Finally, in some of the better watered areas in the lowlands to the north of the Limpopo River, there are small isolated patches of Seasonal Miombo Woodland. The total area covered by this biomass class in Mozambique is about 190,839 km² or 23.9% of the country.

It is an intermediate miombo woodland type between the wet and the dry phases. This relationship is clearly reflected in its geographical distribution in the Zambezi Valley. In northern parts of the valley it is representative of slightly drier conditions than are found in the surrounding wet miombo woodland, and in the south it represents wetter conditions than the surrounding Dry Miombo Woodland and scrubby vegetation. This intermediate nature can also be seen in the phenological characteristics. Photosynthetic activity from January to April is high (NDVI is about 185); these values are similar to the wet miombo woodlands and significantly higher than the Dry Miombo Woodland class. Productivity begins to decline in May, a month earlier than in the wet miombo woodland. Low values of photosynthetic activity occur between July and September. The NDVI values during this period (145–150) are similar to those found in the Dry Miombo Woodland but are much lower than the wet variant. Activity rises sharply from October onwards. This phenological pattern is indicative of areas with substantial moisture reserves in the wet season and a marked dry season drought (either due to rainfall or low moisture holding capacity of soils) which places the woodland under severe moisture stress for up to three months. Consequently, it is semi-deciduous in nature.

Structurally, the woodland is intermediate between the other types of miombo woodland. In less degraded and moister areas the canopy trees reach 8–10 in height. Due to the seasonality, it is more open than wet miombo woodland. The underlying grass layer is better developed and there is a greater proportion of smaller trees and shrubs. In more degraded and drought-prone areas, it grades into an open savannah with a well-developed herb and grass layer and widely spaced trees of up to 8 m tall. The canopy contains similar species to other types of Miombo Woodland and is dominated by *Brachystegia spiciformis* and *Julbernardia globiflora*. The biomass class also includes the extensive tea plantations in Zambezia.

The fuelwood resources of Seasonal Miombo Woodland are relatively high, despite the strong lull in dry season productivity. Where it occurs in association with dry woodland areas, it will form important fuelwood sources; but in wetter regions they will be less important than the surrounding wet miombo woodland.

**Dry Miombo Woodland**

Dry Miombo Woodland is most extensive in Tete Province where it dominates the southerly and central parts of the Zambezi Valley. Related areas are found in northern Manica and Sofala Provinces. Isolated patches are also found in the Limpopo Valley in Gaza Province, but most of these can be classified as Woody Scrub and Thicket. These areas are characterized by low annual rainfall, and in the Zambezi Valley the boundary of the Dry Miombo Woodland closely follows the 800 mm isohyet. A further area of related dry woodland is found in Cabo Delgado Province where it forms a strip parallel to the coast; the rainfall in this region, however, varies from 900 to 1,200 mm. The total area of this biomass class in Mozambique is about 148,545 km² or 18.6% of the country.

Dry Miombo Woodland shows a marked seasonality in photo-synthetic activity. Between January and May, this is high (with NDVI values of about 175) but lower than in the other types of miombo woodland. This falls dramatically in June and July, and from late July until October NDVI values are about 140. This low period of photosynthetic activity in the drier miombo woodlands leads to a deciduous period which can last for up to four months.

Dry Miombo Woodland has a maximum canopy height of about 8 to 10 m. As it becomes drier, it can be restricted to as little as 3 m and takes on an open, shrubby savannah form. It is the latter form which is more commonly encountered (Malleux, 1980). Dry

Miombo Woodland is floristically poor and dominated by *Brachystegia spiciformis*, *B. boehmii* and *Julbernardia globiflora*. Scattered amongst the canopy are smaller trees of *Uapaca* spp., *Monotes* spp. and *Protea* spp. Where the canopy is very disrupted, shrubs commonly invade and a grass layer develops of 0.6 to 1.2 m in height. In these cases, the dominant tree is *B. boehmii* but other small trees and shrubs such as *Burkea africana*, *Faurea speciosa*, *Hymenocardia acida*, *Ochna schweinfurthiana*, *Parinari curatellifolia*, *Swartzia madgascariensis*, *Syzygium guineense* subsp. *guineense*, *Terminalia brachystemma* and *Vangueriopsis lancifolia* invade the canopy.

The fuelwood potential of Dry Miombo Woodland is restricted by two factors. First, the long dry season during which wood production is reduced and, secondly, the low levels of woody biomass (low wood: grass ratio). In many cases, Dry Miombo Woodlands in Mozambique are areas of fuelwood supply shortage.

Thicket-like vegetation is also found in the Dry Miombo Woodland biomass class. Some are very degraded areas whilst others are probably edaphic climax, dry thicket vegetation. This vegetation is common in Gaza and Inhambane Provinces. The proportion of shrubs to trees varies from high-thicket vegetation, in which the trees are dominant and reach 10 m, to low thicket in which shrubs of up to 3 m are more important than the trees (which occasionally reach 8 m). With decreasing height, the canopy becomes more open, and grasses and herbs increase in importance.

The thickets are floristically related to the miombo woodlands and in many cases are a degraded form of them. The main tree species are *Brachystegia spiciformis*, *B. boehmii*, *Julbernardia globiflora*, *Acacia nigresecens* and *Albizia* spp. Some of these species also take on a shrub habit but other shrub species invade the thicket vegetation as well.

Wood and shrub thickets provide readily accessible woody resources in areas with otherwise low potential. Although locally the growing stock is relatively high, overall levels of annual productivity are very low particularly due to severity of the dry season. Consequently, they do not represent a long term woody-biomass resource.

**Mopane Woodland**   There are small patches of dry, seasonal woodland dominated by *Colophospermum mopane* in the Zambezi Valley. They are almost entirely restricted to Tete Province, covering some 11,544 km² or

1.4% of the country. Although similar to the drier miombo woodland types in structure, they have diagnostic floristic and phenological features.

NDVI values in January are similar to those in the Dry Miombo Woodland (about 175 to 180) but they decline slowly from February onwards, reaching about 140 in September and October. This indicates a long drawn out seasonal drought similar to that experienced in Dry Miombo Woodland. The main difference between the two biomass classes lies in the way in which vegetation die-back starts and the rate at which it progresses. In the Mopane Woodland die-back starts much earlier and is more steady than in the Dry Miombo Woodland.

Mopane Woodland in Mozambique is very degraded and structurally similar to Dry Miombo Woodland. It is characterized by a low, very open canopy dominated by *Colophospermum mopane* rather than *Brachystegia* spp. The invasive small trees and shrubs are similar to the Dry Miombo Woodland and there is a well-developed grass and herb layer.

The fuelwood potential of Mopane Woodland is similar to that of Dry Miombo Woodland because of the low growing stock and annual productivity levels. Consequently, Mopane Woodlands have low fuelwood reserves and indicate areas of shortage.

**Coastal Forest Mosaic** Coastal Forests are an important littoral forest zone found in various places along the Mozambique Coast in Inhambane, Nampula, Sofala and Zambezia Provinces. In many areas they form either a narrow fringe of mangroves along the coast, or riparian woodland in the lowest sections of most river systems. Their small, individual areal extent means that they cannot often be identified on 8 km GVI imagery. However, extensive areas can be detected, at the mouth of the Buzi River and around Inhambane. In these two areas, dry evergreen coastal forests are found. Other much smaller areas can be seen along the coast and the lower courses of the major rivers. The total area is 46,908 km² or 5.9% of the country.

Although they are essentially evergreen, these forests have a distinct phenology which separates them from the Evergreen Forest and Woodlands. They show little evidence of seasonality and NDVI values range from about 190 in the wet season to 180 in the dry season. The wet season values are very similar to those for Evergreen and Wet Forests and Woodlands. The main differences lie in the levels of dry-season photosynthesis, particularly in August and September, that are higher than those found in other evergreen vegetation types. Three main species of mangrove (*Rhizophora mucronata, Avicennia marina* and *Sonneratia alba*) are found in close association in areas of mangrove forest.

Riparian forests are found along the major rivers which drain through Mozambique. It is often difficult to distinguish them from the surrounding woodland and forest because of their narrow, restricted occurrences. These forests are best developed on the lower courses and estuaries of the rivers, where there are specific edaphic conditions (alluvial soils and high water-tables with seasonal flooding).

There is, of course, a substantial overlap between the riparian and the adjacent coastal forests. They form a well-structured gallery forest generally with taller trees than are found in the surrounding forests. In areas with high water-tables a swamp forest variant is found, but this is restricted in occurrence and is dominated by *Barringtonia racemosa*.

Estuarine Forests have high levels of woody growing stock and annual productivity. However, this favourable situation for fuelwood resources is subject to problems of accessibility because of daily and seasonal flooding.

## Dry Riparian Woodland

This biomass class is restricted to the lower courses on the Limpopo, Olifantes and Save Rivers. It occurs in Gaza, Inhambane and Maputo Provinces. Unlike the high-productivity riparian forests dentified in the previous biomass class, the vegetation in these river valleys is a low biomass, low productivity, seasonal tree and grass savannah. It has an areal extent of 22,958 km$^2$ or 2.9% of the country.

Dry Riparian Woodland has a relatively high rate of photo-synthetic activity in January, when NDVI values are about 175. They decline slowly throughout the wet season and after May, they decline more rapidly reaching very low levels between August and October, when NDVI values vary between 135 and 145. This indicates a strong seasonal drought and relatively little arboreal vegetation. The vegetation is essentially a grass-dominated savannah with shrubs or small trees, reaching heights of 3–5 and 5–10 m respectively. The species are related to the secondary invasive tree and shrub species in the surrounding woodland. It appears to be a very degraded form of riparian woodland.

Dry Riparian Woodland occurs along many large rivers and consequently, in the south, there are seasonal accessibility problems due to flooding during the wet season. Nevertheless, Dry Riparian Woodland situated alongside rivers running through the drier and degraded woodlands is an important fuelwood resource because of its higher growing stock and levels of productivity.

**Degraded Agricultural Land**

Extensive areas of degraded agricultural land are found in the Changane and Limpopo Valleys in Gaza and Inhambane Provinces. They total about 19,288 km$^2$ or 2.4% of the country. These areas are characterized by low levels of productivity and a marked seasonality. They are dominated by agriculture which has destroyed the vegetation to such an extent that only very small patches of thicket and grass savannah are left between farms. These are usually either areas of very low scrubby vegetation (less than 3 m in height) or open, grass-dominated savannah with isolated trees which reach heights varying from 5 to 10 m. The growing stock is very low and, in the case of the more open savannah, very scattered. The annual productivity levels are also very low. Consequently, the fuelwood potential is severely restricted in these areas.

**Lowland Sublittoral Forest and Bushland**

Lowland Sublittoral Forest and Bushland extends from the South African border northwards to the Olifantes Valley. It is restricted to Maputo Province and covers 3,638 km$^2$, only 0.5% of Mozambique.

The main forest type is sand forest which has a dense 10-25 metre-high thicket with three main strata – a canopy, a sub-canopy with small trees, and a poor ground herb layer. It is mainly found on sandy soils between the littoral zone and the mountains, and degrades into an open deciduous bushland. The canopy of semi-deciduous trees is 5-7 m tall with a well-developed woody understorey. Degraded forms are attributed to damage due to high population densities (Moll and White, 1976); they occur throughout the area.

The fuelwood potential of this biomass class is variable, particularly in terms of growing stock. Despite relatively low levels of annual productivity, it is used extensively for fuelwood in the vicinity of Maputo.

**Lubombo Hills Woodland**

The association of montane grassland and bushland is found in the Lubombo Hills on the Swaziland border. It is entirely restricted to Maputo Province and only covers 1,473 km². The Lubombo Hills are characterized by a savannah of grassland and bushland containing isolated patches of forest. Generally, the fuelwood potential of these hills is very low due to the low growing stock, the low wood : grass ratio, and the very low levels of annual productivity. Consequently, fuelwood potential is severely limited.

**Littoral Grassland**

Littoral Grasslands are most extensive to the south of Maputo but are also found along the coast to the north of the capital in Maputo and Sofala Provinces. The total area covered is 10,043 km², about 1.3% of the country.

These grasslands are edaphically controlled (Moll and White, 1978) but there is a strong possibility that, being so close to Maputo, they can be attributed partially to anthropogenic causes as well. There is evidence to suggest that if left they would revert to a scrub forest or, where more waterlogged soils occur, palm veldt. They form a sward of tussock grasslands, 1–1.5 m tall and provide little in terms of woody biomass, except in the palm veldt. However, even here the fuelwood potential is low because of the low growing stock and inability of the palms to regenerate under sustained pressure.

**BIOMASS SUPPLY**

Mozambique has the second largest level of growing stock of all SADCC member states, slightly more than Tanzania or Zambia. The MAI outlook for the country is, if anything, slightly better. The total growing stock of 3515.3 million tonnes and MAI of 114 million tonnes when divided by the population, gives values of 284.4 tonnes/person and 9.2 tonnes/person respectively (see Table 7.1).

Levels of woody-biomass supplies are favourable in most provinces, but particularly so in Nampula, Niassa and Zambezia, which between them hold 50% of the growing stocks and 44% of the total MAI (see Table 7.2). The situation in other provinces is more variable. For instance, in the six provinces with areas ranging from approximately 61,000 to 82,000 km², Nampula alone has 13% of the growing stocks and 12.4% of the MAI. The others have lower proportions of both. They are Inhambane (with 9.7% of growing stock and 15.2% of the MAI); Sofala (with 8.8% and 8.2%); Cabo

**Table 7.1** MOZAMBIQUE: SUMMARY OF GROWING STOCK AND MAI

| Biomass Class | Area | | Growing Stock | | MAI | |
|---|---|---|---|---|---|---|
| | (km²) | (%) | (mill t) | (%) | (mill t) | (%) |
| Evergreen Miombo Woodland and Coastal Forest | 121,002 | 15.1 | 862.0 | 24.5 | 27.2 | 23.8 |
| Wet Seasonal Forest and Woodland | 223,141 | 27.9 | 1,589.2 | 45.3 | 50.2 | 44.0 |
| Seasonal Miombo Woodland | 190,839 | 23.9 | 378.8 | 10.8 | 9.4 | 8.3 |
| Dry Miombo Woodland | 148,545 | 18.6 | 140.2 | 4.0 | 3.6 | 3.2 |
| Mopane Woodland | 11,544 | 1.4 | 42.7 | 1.2 | 1.3 | 1.1 |
| Coastal Forest Mosaic | 46,908 | 5.9 | 416.4 | 11.8 | 19.0 | 16.7 |
| Dry Riparian Woodland | 22,958 | 2.9 | 26.3 | 0.7 | 1.2 | 1.0 |
| Degraded Agricultural Land | 19,288 | 2.4 | 45.1 | 1.3 | 1.6 | 1.4 |
| Lowland Sublittoral Forest and Bushland | 3,638 | 0.5 | 3.4 | 0.1 | 0.1 | 0.1 |
| Lubombo Hills Woodland | 1,473 | <0.1 | 11.2 | 0.3 | 0.4 | 0.4 |
| Littoral Grassland | 10,043 | 1.3 | 0 | 0 | 0 | 0 |
| TOTAL | 799,379 | | 3,515.3 | | 114.0 | |

**Table 7.2** SUMMARY OF GROWING STOCK AND MAI BY PROVINCE

| Province | Proportion of: | |
|---|---|---|
| | Growing Stock (%) | MAI (%) |
| Cabo Delgado | 7.7 | 7.0 |
| Gaza | 6.5 | 4.4 |
| Inhambane | 9.7 | 15.2 |
| Manica | 7.7 | 7.2 |
| Maputo | 2.4 | 5.9 |
| Nampula | 13.0 | 14.4 |
| Niassa | 18.2 | 17.1 |
| Sofala | 8.8 | 8.2 |
| Tete | 7.0 | 6.1 |
| Zambezia | 19.0 | 14.5 |

Delgado and Manica (with 7.7% each of the growing stock, and 7.0% and 7.2% of the MAI); and finally Gaza (with only 6.5% of the growing stock and 4.4% of the MAI). Maputo Province is much smaller, only 26,358 km², but it only has 2.4% of the growing stock and 5.9% of the MAI. Another province with supply problems is Tete. It is the third largest province but only has 7% and 6.1% of the growing stock and MAI respectively.

The inequitable resource situation is caused by the large growing stocks (2,451.2 million tonnes) and MAIs (77.4 million tonnes) in the two densest miombo woodland and forest classes. Despite the fact that they only cover 43.1% of Mozambique, they contain 69.8% of the growing stock for the country and a similar proportion of the MAI. The only other biomass class which is over-represented is the Coastal Forest Mosaic (see Table 7.1).

The drier miombo and savannah woodlands, which cover almost half of Mozambique (47.3%), only possess 16.8% and 13.6% of the growing stock and MAI respectively. A similar situation is also found in the Degraded Agricultural Land and Littoral Grassland classes.

Such a polarization of woody-biomass resources inevitably creates a division between provinces with low potential and high potential woody-biomass supply. Those with low supply are:

- Gaza
- Maputo
- Tete

A further group can be identified in which excessive exploitation would easily lead to supply shortfalls. These are:

- Cabo Delgado
- Manica

In fact, in north-west Cabo Delgado there is already a large area with low supply potential.

The discussion of woody-biomass resources in Mozambique would not be complete without referring to the ongoing civil war and how it affects them. First, in the areas of intensive guerilla acivitity there has been an exodus of the local people. Extrapolated over long periods, this may lead to the re-establishment of woody biomass in intensively cultivated areas. This may happen in Niassa and Tete Provinces, where many refugees are leaving for Malawi, but in the south (Gaza, Inhambane and Maputo) the recurrent droughts militate against rapid vegetation recovery. In well-wooded areas, the war has little affect on the actual resource base because the woodlands are resilient and very extensive. However, guerilla activity restricts people to the vicinity of "safe" villages, putting pressure on the immediate woody-biomass resource base. This type of localized deforestation is too small to see using AVHRR GVI data. A further affect of the war is that the access restrictions to forest reserves and national parks have become almost redundant; therefore they have not been considered.

# 8.    Swaziland

**GENERAL DESCRIPTION AND BIOMASS CLASSES**

Swaziland is divided into four main regions. From west to east across the country these are the Highveld, the Middleveld, the hot dry plains of the Lowveld, and the Lubombo Hills on the Mozambique border. This division was used by I'Ons (1967) to map the following nine veld types in the country:

Highveld – Mountain Sourveld
– Highland Sourveld

Middleveld – Upland Tall Grassveld
– Moist Tall Grassveld
– Tall Grassveld
– Dry Tall Grassveld
– Upper broadleaved tree savannah with Hillside Bush

Lowveld – Lower broadleaved Tree Savannah
– Dry *Acacia* Savannah

Swaziland has been divided into eight biomass classes in this assessment. The main factors used in the identification of these biomass classes were total biomass and productivity. The biomass classes represent land-use types in Swaziland rather than vegetation types, as is the case in a number of the other countries in the SADCC region. This is not entirely unexpected because of the large amount of land under forestry and agriculture. The biomass classes range from those that are highly productive with a large total biomass (e.g., forestry and irrigated cash-cropping) to those that have much lower biomass and limited productivity (e.g., rainfed cultivation and overgrazed grasslands and savannahs).

Both productivity and biomass are important indicators of

fuelwood potential. Generally, it could be assumed that the higher the productivity and biomass, the greater the fuelwood resources. However, this is not the case in Swaziland for the following two reasons. First, much of the land in the most productive biomass classes is given over to forest reserves, plantations and irrigation schemes; in all three cases, access is limited. Secondly, although the irrigation schemes are indicative of high levels of biomass and productivity, the amount of woody material is extremely low, if there is any at all. Consequently, most of the accessible fuelwood resources in Swaziland are found in areas with moderate levels of biomass and productivity (the areas of Dense Bushland and Woodland). These are mainly situated in two districts:

- Lubombo
- Shishelweni

Areas of fuelwood scarcity exist for two reasons. First, because of restricted access to high-potential wood reserves. This occurs mainly within the following districts:

- Hhohho
- Manzini
- Shishelweni

Secondly, in areas with very low levels of biomass and productivity, this can either be a natural phenomenon or, as is common in Swaziland, due to overgrazing and soil erosion. This is mainly a problem in parts of the following districts:

- Hhohho
- Lubombo
- Manzini

If these factors (i.e., accessibility and the level of potential) are examined together, the four districts can be ranked in order of fuelwood potential as follows:

- Shishelweni (highest potential supply)
- Lubombo
- Manzini
- Hhohho (lowest potential supply)

The eight biomass classes (see map on p.125) were identified from NOAA-7 LAC (4 x 4 km) data and by referring to previous work. Particularly useful here was the 1:250,000 Satellite Map of

Swaziland compiled by ITC (Netherlands) and Murdoch's (1968) comprehensive study of land capability. Field-checking of image interpretations and growing-stock estimates was carried out in Swaziland in August 1986. The biomass classes are:

- Highveld Forest
- Dense Plantation Stands
- Wattle and Eucalyptus Plantations
- Lubombo Hills Woodland
- Dense Bushland and Woodland
- Sparse Bushland and Woodland
- Open Grassy Savannah
- Irrigated Agriculture

## DESCRIPTION OF INDIVIDUAL BIOMASS CLASSES

### Dense Plantation Stands

Forest Plantations are found extensively in the western Swaziland Highveld; particularly in these three main areas:

- Piggs Peak, Hhohho District
- Usutu, Manzini District
- around Nhlangano, Shishelweni District

The densest forest stands can be identified within the surrounding plantations and reserved forests in all three areas by the fact that they have higher NDVI values than the surrounding forests. They account for 1,049 km² or 6.0% of the country.

The coniferous plantations (Piggs Peak and Usutu) are dominated by pines, in particular Caribbean imports such as *P. elliottii*, *P. patula* and *P. taeda*. Most of the wood from these forests is destined for South African pulp mills. The southern plantations in Shishelweni District are mainly gum (especially *Eucalyptus grandis* and *E. saligna*) and black wattle (*Acacia mearnsii*). These are dealt with in more detail in the Wattle and Eucalyptus Plantation class.

These areas represent the forest stands which have a combination of relatively high levels of biomass and high annual growth. The distribution of these areas within the overall scheme of the plantations will depend on the stage of growth of different forest plots and also on felling schedules. Therefore, it will vary over time. More detailed information regarding the forestry operations, including an assessment of their fuelwood potential, is found in the Highveld Forest biomass class. Although wood production and

**Key to facing page:**

K Middleveld/Highveld Dense Woodland and Bushland
*K* Lowveld Dense Woodland and Bushland
M Middleveld/Highveld Sparse Woodland and Bushland/Open Grassy Savannah
*M* Lowveld Sparse Woodland and Bushland/Open Grassy Savannah
Q Lubombo Hills Woodland
W Highveld Forest
X Dense Plantation Stands
Y Wattle and *Eucalyptus* Plantations
Z Irrigated Agriculture

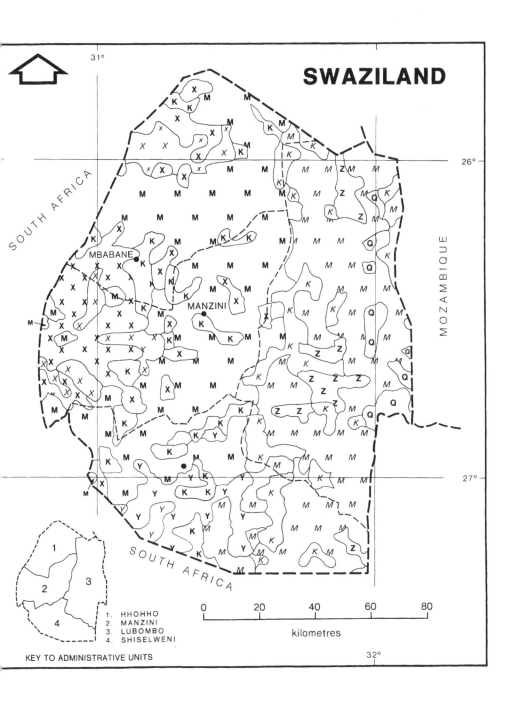

# SWAZILAND

N

31°

26°

27°

32°

SOUTH AFRICA

MOZAMBIQUE

SOUTH AFRICA

MBABANE

MANZINI

## KEY TO ADMINISTRATIVE UNITS

1. HHOHHO
2. MANZINI
3. LUBOMBO
4. SHISELWENI

0  20  40  60  80

kilometres

biomass is high in these stands, they represent commercial timber operations and access to fuelwood is extremely limited.

**Highveld Forest**

Forests in the Highveld fall into two groups. First, plantations which are found extensively in the western Swaziland Highveld, particularly in these three main areas:

- Piggs Peak, Hhohho District
- Usutu Forests, Manzini District
- Shishelweni District

Secondly, natural forests, most of which are now reserved in the northern and central Highveld. These two forest types account for 2,466 km² or 14.2% of the country.

The Piggs Peak and Usutu plantations are dominated by Carribean pines such as *P. elliottii, P. patula* and *P. taeda* which are destined for pulp mills in South Africa. Murdoch (1968) suggested that, on the basis of suitable soil types, about 4,050 km² of the Highveld and cool Middleveld were suitable for pine and gum afforestaton (i.e., areas with mean annual rainfalls greater than 1,000 mm and 890 mm for the two tree types respectively). He noted further that food-crop production on the Middleveld was very important and that forestry–agriculture conflicts could occur. He also estimated that there was about 1,460 km² of Highveld with a mean annual rainfall above 1,150 mm, making it suitable for pine afforestation. In the late 1960s only 120 km² was under pine forest, but there has been significant afforestation since then.

The coniferous plantation area has expanded steadily since the first areal estimates in 1950 (see Table 8.1). Timber statistics for 1982 (Government of Swaziland, 1983) show that the ratio of planted to natural forest is 45.5 : 1 (see Table 8.2) underlining the dominance of Swazi woodlands by plantations. Two-thirds of the plantations are company owned.

Although Compton (1966) suggested that there was never any natural forest on the Highveld, 7% of the Highveld (about 364 km²) supports a forest dominated by *Cussonia umbellifera, Podocarpus latifolius, Rawsonia lucida* and *Xymalos monospora* in sheltered hollows and ravines. This is almost certainly natural Highveld Forest which is now reserved in the Swazi National Forest. Field observations of one forest plot in 1986 showed dense woodland with a canopy reaching a height of about 12 to 15 m, and a lower under-storey of

**Table 8.1** SWAZILAND: WOODLAND AND
FOREST AREAS

*Forest type and area (km²)*

| Year | Pines | Gum (E. saligna) | Poplar (P. deltoides) | Wattle (A. mearnsii) | Indigenous |
|------|-------|------------------|------------------------|----------------------|------------|
| 1950 | 67.2 | 9.0 | 0 | 80.6 | 291.2 |
| 1960 | 313.6 | 33.6 | 2.2 | 76.2 | 224.0 |
| 1964 | 385.3 | 40.3 | 2.2 | 65.0 | 224.0 |
| 1967 | 403.2 | 58.2 | 2.2 | 51.5 | 224.0 |
| 1981 | 747.3ˣ | 215.6ʸ | – | 26.4 | – |
| 1982 | 757.4ˣ | 198.6ʸ | – | 26.6 | – |
| 1984 | 1,049.4 | | 63.9 | – | ᶻ |

ˣ coniferous plantations
ʸ incl. *E. grandis*
ᶻ incl. with coniferous plantations
– no data

Data from Murdoch (1968), Government of Swaziland 1982 and this work.

tree species with heights varying from 4 to 9 m. The woodland is dense with about 3,400 trees/ha.

Woody-biomass resources are potentially high in this class. Unfortunately, however, almost all of the forest cover in this class is either reserved or managed under plantation schemes, making access severely limited.

**Irrigated Agriculture**  Irrigated agriculture is found throughout the Lowveld but the largest and most productive areas are included in this biomass class. These are the Big Bend and Vuvulane Farms in Lubombo District. In total they cover 1,465 km² or 8.4% of the country.

Cash-cropping under irrigation is an important land-use category in parts of the Lowveld. Where it occurs, it is locally dominant and accounts for all of the land. The main areas of irrigated agriculture (with estimated sizes from Murdoch, 1968) are all included in this biomass class. They are:

**Table 8.2** AREA OF PLANTATIONS BY LAND-USE AND OWNERSHIP TYPE ON DECEMBER 1982

| Land-Use Category | Swaziland Ha | % To Total | Individual and Partnership Ha | % To Total | Registered Companies Ha | %To Total | Other[1] Ha | %To Total |
|---|---|---|---|---|---|---|---|---|
| A. Forest: | 106,914 | 57.8 | 7,817 | 26.3 | 97,612 | 67.0 | 1,485 | 15.5 |
| Man-Made | 100,916 | 54.6 | 7,393 | 24.9 | 92,553 | 63.5 | 970 | 10.1 |
| Natural | 2,231 | 1.2 | 190 | 0.6 | 1,707 | 1.2 | 334 | 3.5 |
| Temporarily unplanted | 3,767 | 2.0 | 234 | 0.8 | 3,352 | 2.3 | 181 | 1.9 |
| B. Forest Servings | 14,159 | 7.7 | 423 | 1.4 | 13,726 | 9.4 | – | – |
| C. All Other Land: Including cultivated, grazing and unused land | 63,912 | 34.5 | 21,504 | 72.3 | 34,328 | 23.6 | 8,080 | 84.5 |
| D. Total Area: Covered by the Census | 184,985 | 100.0 | 29,744 | 100.0 | 145,666 | 100.0 | 9,565 | 100.0 |

Source: Govt. of Swaziland, 1982
[1]Including educational and religious institutions

- Sugar cane at Big Bend, Mhlume and Lubombo
- Rice (28.3 km²)
- Citrus (24.3 km²)
- Mixed sugar cane and vegetable (28.3 km²) at Vuvulane
- Other crops, mainly maize, wheat, cotton and vegetables (812 km²)

The area under irrigation has expanded since the late 1960s, although it is still concentrated in the same Lowveld areas.

In this class, woody-biomass resources are severely limited as most of the land cover is given over to crops or infrastructure (roads, factories and housing).

**Wattle and Eucalyptus Plantations**

Wattle and Eucalyptus Plantations are concentrated within Shishelweni District, particularly around the towns of Hlatikulu and Nhlangano. They account for only 163 km², a mere 0.9% of the country.

The southern plantations are mainly of gum (*Eucalyptus grandis* and *E. saligna*), poplar (*Populus deltoides*) and wattle (*Acacia mearnsii*). The estimated areas from 1950 to the present-time can be found in Table 8.1. Black wattle was once an important tree crop in the area, its bark being stripped for tannin extraction and its wood used for mining timber. Both products were shipped to South Africa. As the tannin demand has declined so has the afforested area. The plantations are left unattended and have become "wattle jungles" (David Gwaitta-Mgumba, personal communication, 1986). The *Eucalyptus* plantations have expanded during the same period; the wood is used mainly for poles and mining timber (Krohn, 1977).

Woody-biomass growing stocks and mean annual production levels are high in these plantations. Access to most, however, is severely restricted, the exception being the "wattle jungles" which provide a valuable local wood resource.

## Dense Bushland and Woodland

Areas of mixed bushland and woodland occur extensively throughout Swaziland. They can be divided in terms of phenology and biomass into "dense" and "sparse" categories and these have been dealt with as two separate biomass classes. A further division has been made on the map (see p. 125) which is related to the topographic zone that the bushland and woodland occurs in (i.e., Middleveld or Lowveld). This division was not evident in the imagery but is based on fieldwork undertaken in 1986.

Dense Bushland and Woodland is mainly restricted to the southern and central Lowveld (Lubumbo District) and the southern Middleveld (Shishelweni District). These scattered patches relate to undisturbed areas of natural bushland and woodland in the Middleveld, and severe overgrazing in both the Lowveld and Middleveld. It is slightly more extensive than Sparse Bushland and Woodland, the total area being 3,308 km$^2$ or 19.1% of the country.

The natural vegetation in the Middleveld is a grassveld. This is generally an open, grassy savannah with clumps of bushes and smaller trees reaching between 5 and 8 m in height but with emergents of up to 15 m tall. There are also areas of naturally dense, woody vegetation which are often the result of soil and hydrological factors. The dominant trees and shrubs in the moister grassland, found on the western parts of the Middleveld, are *Acacia karroo*, *Maesa lanceolata*, *Syzygium cordatum* and *Vernonia ampla*. In the drier grasslands in the eastern Middleveld the main arboreal components are *Acacia nilotica, Dichrostachys cinerea, Ficus capensis, Maytenus senegalensis* and *Sclerocarya birrea*; here the woodland is usually between 4 and 5 m in height. Access to the naturally dense woodlands is now very restricted as the Middleveld is densely populated; it contains 41% of the population (Krohn, 1977) and is extensively cultivated and grazed.

The natural vegetation of the Swaziland Lowveld is grassland but, as a result of overgrazing, there has been an invasion of thorny, bushy trees. This ecological situation is commonly found on granitic and basaltic plains in southern Africa and is known as the "*Acacia nigrescens* Tropical Plains Thornveld". It is found on the Swaziland Lowveld in areas which are neither irrigated nor extensively grazed. The areas of the Lowveld that fall into this class are the areas where bush vegetation is most dense, indicating the heaviest levels of overgrazing in the past or relict riparian woodland. The trees and bushes are highest in these areas, sometimes reaching up to 15 m. These thickets are dominated by *Acacia nigrescens, Combretum zeyheri, Dichrostachys cinerea* subsp. *africana* and *Sclerocarya caffra*. The tree density in sampled plots was variable, ranging from 350 to 3,200 trees/ha; more usually it was between 1,000 and 1,300 trees/ha.

Dense Bushland and Woodland has a relatively large growing stock but productivity is limited. Nevertheless, this class represents one of only two biomass classes in Swaziland (the other being Sparse Bushland and Woodland) with wood resources which are physically accessible and have not been extensively reserved. Consequently, they represent an important fuelwood biomass class in Swaziland at the present time.

**Sparse Bushland and Woodland**

Sparse Bushland and Woodland occurs throughout the Middleveld (particularly in the south) and the Lowveld. There are also smaller areas around the edges of the Highveld Forests and Plantations and in the Lubombo Hills on the Mozambique border. It is most common in Hhohho and Lubumbo Districts.

This class occurs in scattered patches throughout Swaziland and usually represents moderately disturbed areas of natural bushland and woodland. On the Lowveld it is indicative of overgrazing. It is less extensive than Dense Bushland and Woodland, only accounting for 2,313 km² or 13.3% of the country.

The grassveld of the Middleveld generally takes the form of an open, grassy savannah with areas of bushes and smaller trees. Observations from field plots (1986) suggest it rarely exceeds 3.5 m in height and the tree density ranges from 250 to 325 trees/ha. In the moister grassland the dominant trees and shrubs are *Acacia karroo, Maesa lanceolata, Syzygium cordatum* and *Vernonia ampla*. In the drier grasslands to the east, the main trees are *Acacia nilotica, Dichrostachys cinerea, Ficus capensis, Maytenus senegalensis* and *Sclerocarya birrea*. Areas of bushes and trees are now very restricted due to cultivation and grazing. The main areas of Middleveld with sparse bushland and woodland are:

- in the north on the South African border
- between Mbabane and Manzini
- on the Lulongweni and Sinceni Hills in the south

The latter area, mainly in Shishelweni District, is the most extensive.

Grassland is the natural vegetation of the Swaziland Lowveld but with overgrazing, thorn trees have invaded and formed the "*Acacia nigrescens* Tropical Plains Thornveld". The main areas of sparse Lowveld bushland and woodland are in Mahuku, Qomintaba and Sigcaweni. In these bushy savannahs, the trees and bushes are interspersed with grasses. The main trees are *Acacia nigrescens* with secondary amounts of *Combretum zeyheri, Dichrostachys cinerea* subsp. *africana* and *Sclerocarya birrea*. The trees can reach up to 10 m. Between them are areas of low bushes, the most common being *Cordia gharaf, Ormocarpum trichocarpum* and *Sclerocarya caffra*. These frequently do not exceed heights of 4 m. The tree density is higher than the Middleveld equivalent, ranging from 325 to 625 trees/ha. The distribution of trees and bushes is strongly controlled by soils, the main relationships being:

| | |
|---|---|
| *Acacia xanthophleoa* | Calcareous clays |
| *A. gillettiae* (thickets) | Calcareous clays |
| *A. nigrescens* | Shallow soils on basic rocks |
| *A. gerrardii* | Clay soils |
| *Dalbergia melanoxylon* | Moderately dry conditions |
| *Spirostachys africana* | Iron or clay pan soils |
| *Zizyphus mucronata* | Iron or clay pan soils |
| *Terminalia sericea* | Light textured soils |
| *Pterocarpus angolensis* | Light textured soils |
| *Strychnos innocua* | Light textured soils |

| | |
|---|---|
| *Albizia harveyi* | Heavy basaltic soils |
| *Lonchocarpus capassa* | Heavy basaltic soils |
| *Maerua parvifolia* | Dry soils |

Eradication of this thorny bushland within the savannah is difficult and expensive and, as the area has been extensively overgrazed, bushy savannah is widespread and is extending at the present time.

On the Highveld there is very little bushland. It only occurs in isolated patches, usually around the fringes of the plantations and woodlands.

**Lubombo Hills Woodland**

Lubombo Hills Woodland is now restricted to a few isolated areas along the Lubombo Hills Escarpment. Most of the woodland has now been cleared for cultivation and it only accounts for 64 km² (less than 1% of the country). It is restricted entirely to Lubombo District.

The natural vegetation on the Lubombo Hills is a very open savannah of small trees and bushes which reach heights of about 3 m. However, in some places on the escarpment there are still dense woodland stands of up to 8 m in height dominated by *Androstachys johnsonii*. These are potentially very important as they have the highest woody-biomass growing stocks in Swaziland outside the Highveld Forests and Plantations. Work by Coetzee and Nel (1977) suggests the Lubombo Mountains *Androstachys* Woodland can be divided into four ecological types, all with different fuelwood potentials. In order of size of growing stocks these are:

- i) dense stands of about 1,400 trees/ha, with trees 4–8 m tall and interlocking canopies;
- ii) moderately dry shrub communities on the summits, with 1,000–1,700 trees/ha varying from 3 to 5 m in height;
- iii) dry shrub communities, with 650 trees/ha varying between 2 and 3 m tall on the summits; *and*
- iv) a mixture of grass and low trees on boulder outcrops.

The immediate fuelwood potential of the remaining areas of woodland is very high but, because of their severely restricted occurrence, they can only be seen as a resource in the short-term.

**Open Grassy Savannah**

Open Grassy Savannah is found mainly in northern Swaziland. In some areas it represents very open grassland with a few bushes (which may be either natural or a result of overgrazing) or areas of rainfed agriculture with fallow grassland and bushland. It occurs mainly to the south and east of the Piggs Peak Plantatations, in the northern Middleveld and Lowveld, and in restricted areas of the Lubombo Hills. It is the most extensive biomass class in Swaziland covering 6,536 km² or 37.7% of the country. It is quite similar to the Sparse Bushland and Woodland in terms of growing stock and productivity and has been included with that class on map p. 125.

In the Highveld, this biomass class represents most of the land which is neither reserved forest nor under plantations, and has a short grass cover which is known as "sourveld" because it is unpalatable to cattle in winter. Krohn (1977) estimates that 29% of the Highveld is grassland and the estimates made from this study are in general agreement. In the Middleveld and, to a lesser extent, the Lowveld, this class represents areas of indigenous Swazi agriculture. The main crops are maize, sorghum, and beans; although other food crops and cash crops (such as avocados, cotton and pineapples) are also grown.

Areas for dryland cropping were investigated by Murdoch (1968) who found that 61.2% of the country (10,640 km²) was suitable for arable agriculture. However, the actual amount under cultivation in the late 1960s was just over 1,000 km² and this was mainly concentrated in the Middleveld and Lowveld. In the Middleveld the vegetation is heterogeneous, but in many areas this biomass class is dominated by rainfed cultivation and grazing. There has been some overgrazing which has reduced the grass cover and bush canopy. As a consequence, grasses have become "sour" and soil erosion has increased. Both of these factors suppress bush regeneration.

Therefore, in the Highveld and Middleveld this biomass class represents areas of overgrazed and overcultivated land. This has led to a reduction in the woody-biomass component (trees and bushes) and an increase in grass species. Consequent upon this reduction in the wood : grass ratio is a lack of fuelwood resources.

In the Lowveld and parts of the Middleveld, the natural vegetation is grassland with scattered bushes. Here, this biomass class represents areas which are probably grazed but not yet overgrazed and, consequently, have not been degraded to bushland as a result of bush encroachment.

The natural vegetation of the Lubombo Hills consists of open savannah with clumps of trees and bushes that are floristically related to the Middleveld, and the eastern coastal plain of

Mozambique and South Africa. This biomass class therefore represents this natural open savannah in this area. It should be noted that areas of dense woodland on the Lubombo Hills are included in the Lubombo Hills Woodland class. The main tree and bush species in this open savannah are *Acacia* spp., *Boscia albitrunca*, *Colophospermum mopane*, *Combretum apiculatum*, *Commiphora glandulosa*, *Diospyros dichrophylla*, *Encephalartos lebombensis*, *Kirkia acuminata*, *Nuxia oppositifolia*, *Pterocarpus rotundifolius*, *Terminalia prunioides*, *Scolopia mundi* and *Ximenia americana*.

Fuelwood resources in the areas of the Lowveld and Lubombo Hills that fall in this biomass class are limited by a low growing stock and a low level of annual productivity.

**BIOMASS SUPPLY**

The total biomass resources of Swaziland are 24.45 million tonnes (see Table 8.3), notionally providing 1,405 tonnes/km$^2$ with 33.3 tonnes per person (based on 1986 population estimates). However these favourable estimates belie the skewed distribution of, and access to, the woody-biomass resources and their low productivity which averages 1.2 tonnes per person per year.

The distribution of growing stock is divided almost equally between the four districts. It varies from 21.5% in Shishelweni to

**Table 8.3**   SWAZILAND: SUMMARY OF GROWING STOCK AND MAI

|  | Area (km²) | % | Growing Stock (mill t) | % | MAI (mill t) | % |
|---|---|---|---|---|---|---|
| Highveld Forest | 2,466 | 14.2 | 3.5 | 14.1 | 0.03 | 2.3 |
| Dense Plantation Stands | 1,049 | 6.0 | 4.1 | 17.0 | 0.54 | 40.6 |
| Lubombo Hills Woodland | 64 | 0.4 | 0.5 | 2.0 | 0.02 | 1.5 |
| Wattle and Eucalyptus Plantations | 163 | 0.9 | 0.7 | 2.8 | 0.14 | 10.5 |
| Dense Woodland and Bushland | 3,308 | 19.1 | 7.7 | 31.5 | 0.27 | 20.3 |
| Sparse Woodland and Bushland | 2,313 | 13.3 | 3.3 | 13.5 | 0.14 | 10.5 |
| Open Grassy Savannah | 6,536 | 37.7 | 4.6 | 19.0 | 0.19 | 14.3 |
| Irrigated Agriculture | 1,465 | 8.4 | 0 | 0 | 0 | 0 |
| TOTAL | 17,364 | | 24.4 | | 1.33 | |

**Table 8.4** SWAZILAND: DISTRIBUTION OF GROWING STOCK AND MAI BY DISTRICT

| Region | Proportion of | |
| | Growing Stock (%) | MAI (%) |
| --- | --- | --- |
| Hhohho | 28.0 | 40.2 |
| Manzini | 23.4 | 11.6 |
| Lubombo | 27.2 | 22.3 |
| Shishelweni | 21.4 | 25.9 |

28.0% in Hhohho (see Table 8.4). This does not take into account the fact that the area of Lubombo District is approximately 55–70% bigger than the other three districts, which are of almost equal area. However, productivity data is less equally distributed, with 40.2% occurring in Hhohho District and only 11.6% in Manzini District (see Table 8.4). From these data it can be concluded that Lubombo has a significantly lower growing stock and MAI on an areal basis, and the MAI is low in Manzini. The district with the highest woody-biomass potential is undoubtedly Hhohho.

The distribution of woody-biomass resources is clouded by the extremely skewed distribution of resources in the eight biomass classes. Four of these classes have growing stock and MAI proportions far in excess of the area of the country that they cover (see Table 8.3). Almost a third of the woody-biomass reserves and a little over 20% of the MAI are found in the Dense Woodland and Bushland Class, although it only covers 19% of the country. The other three classes falling into this group take up small areas. They are Dense Plantation Stands with 6.1% of the country but 17% of the growing stock and 30.4% of the MAI; the Wattle Plantations in Shishelweni District with 0.9% of the country but 2.8% of the growing stock and 12.5% of the MAI; and Lubombo Hills Woodland with 0.4% of the country but 2% of the growing stock and 1.8% of the MAI.

Three areally extensive classes – Open Grassy Savannah 6,536 km², Highveld Forest (2,466 km²) and Irrigated Agriculture (1,465 km²) – account for 60.3% of the land area, and only 33.2% of the growing stock and 15.2% of the MAI. What is particularly critical to woody-biomass resource provision is that the largest biomass class, the Open Grassy Savannah, covers over a third of the country but accounts for less than 19% of the woody-biomass resources. The Sparse Woodland and Bushland Class has almost equal proportions of land cover and resources.

The distribution of biomass supply data by class needs to be examined alongside land tenure data. Four of the biomass classes – Highveld Forest, Dense Plantation Stands, Wattle and Eucalyptus Plantations and Irrigated Agriculture – are almost entirely privately owned. In the first three classes, access to woody-biomass resources is severely restricted. This is either because of the extensive woody-biomass demands of commercial timber exploitation. In the irrigated areas there are few woody biomass resources.

These four classes cover 5,143.9 km² or 29.5% of the country. They account for only about the same proportion of growing stock (34%) but for 45.6% of the MAI. This means that the relatively more accessible open savannahs, degraded savannahs with bush encroachment and isolated patches of denser woodland, hold 66% of the woody-biomass resources in 70.5% of the country. Less encouraging though is the fact that this woody biomass grows slowly and only a little over half (54.4%) of the MAI is held in the accessible two-thirds of Swaziland.

If the complexities of the Swaziland land-tenure situation are examined in more detail, it can be seen that further restrictions are placed on the more accessible woody-biomass resources of the savannah and bushlands. The land tenure situation in such areas is complex and various situations exist; including game reserves with no access for fuelwood collection. The largest of these is the Hlanwe Game Sanctuary, which encloses about 290 km² of Dense Woodland and Bushland. Other wildlife reservations are found in Hhohho District three of which – Hawane, Malototja and Mlilwane – enclose about 235 km² of Sparse Woodland and Bushland.

Much land is used for dryland agriculture – mainly rainfed maize cultivation and cattle grazing. This land is held under the two broad tenure groupings described below.

i) Some land is privately owned by expatriate or Swazi farmers. These are usually large farms with a variety of rangeland types varying from good, open, grassy savannah to areas that have been abandoned due to bush encroachment. However, population and stock pressure on these farms is generally low and they tend to have quite good rangelands with minimal bush enroachment. This is good from the viewpoint of woody-biomass resource supply because rangeland degradation, due to bush encroachment after overgrazing, leads to an increase in the available woody-biomass resources on the rangeland.

ii) The situation is more serious on the Swazi National Land (SNL) which, in many areas, suffers from overpopulation and large herd sizes. This leads to range degradation and bush encroachment. At the same time, the rangeland deteriorates and the woody-biomass resources increase. This simple argument, however, needs to be applied carefully in Swaziland. The SNL is divided into a series of chiefdoms, some of which are already overpopulated and show signs of severe range deterioration. Others are underpopulated and still have extensive areas of good open rangeland and dense (natural?) woodland.

All of the chiefdoms are under pressure, at the present time, to provide land for an increasing rural population. This is because migrant labourers have returned from South Africa and are going back to the land. It is also due to government policy which discourages settlement in the main towns of Mbabane and Manzini. The influx of these migrants into each chiefdom varies and creates differential pressures on woody-biomass resources. There is also further pressure on the SNL held under the chiefdom system, due to consolidation of chiefdom land into dryland cultivation areas, range and settlement zones. This often dislocates the people from the accessible woody-biomass resources on the range. In the central and southern lowveld, extensive woodland and bushland clearance is occurring on SNL for cotton cultivation.

Swaziland suffers a significant woody-biomass resource supply problem. The most productive classes, in terms of woody-biomass supply, are inaccessible for most of the population. More accessible land covers about two-thirds of the country and holds about the same proportion of the total woody-biomass growing stock, although its overall level of productivity is generally low. This restricts extensive exploitation. However, the distribution of better woody-biomass resources in these areas revolves around existing

woodlands with low amounts of disturbance, and degraded rangeland with bush encroachment. This pattern of woody biomass is
intricately linked to the complex land-tenure situation, which in
turn conditions access to the resources.

# 9.  Tanzania

GENERAL
DESCRIPTION AND
BIOMASS CLASSES

The vegetation of Tanzania comprises a wide range of types including tall closed forests, many kinds of woodland, open grassland, and areas of semi-desert. The closed canopy forests occur in very restricted areas and are mostly limited to uplands and coastal lowlands. However, the most common types of vegetation are the miombo woodlands and related wooded grasslands. The closed forests and denser miombo woodlands account for 41.2% of the land area or about 370,000 km². In contrast, the drier woodlands and wooded grasslands cover approximately 186,700 km² and account for 22.2% of the land area.

These areal estimates of miombo woodlands and related wooded grasslands are based on 1984 satellite imagery and equate well with Tanzanian Government estimates made in 1983 (Forest Division, 1984) of 431,870 km² of the savannah woodland; and Lanly's 1981 estimate of 420,400 km² of all types of woodland (closed forest and savannah woodland). The differences between the estimates made in this study and those of the Tanzanian government are due to the inclusion of tall, closed forests and some of the sparser wooded grassland areas in the forest classes. The difference between Lanly's 1981 estimate and those made here are due to the inclusion of wooded grasslands.

Many woodland and forest resources are under pressure in Tanzania and Lanly (1983) has estimated a mean annual deforestation rate of 10,000 ha/yr. This is similar to that of Mozambique but lower than those of Angola and Zambia. These woodlands and forests are essentially restricted to the west and east of the country. The western woodlands and forests are found on the Western Plateau and the mountains associated with the Rift Valley Escarpment. The eastern woodlands and forests are associated with the Coastal Lowlands, South-East Plateau and the southern zone of mountains (Berry, 1971). These two large, wooded areas are divided by a large triangular-shaped area, with its apex pointing south-west towards Malawi, of semi-arid bushland and thicket vegetation. In total it covers 292,419 km$^2$, a little over a third of the country (34.8%). It is found in central Tanzania on the eastern part of the western Plateau, the Masai and Handeni Plateaux, and the Central and North-East Highlands (Berry, 1971). It is almost entirely restricted to areas with a mean annual rainfall of less than 800 mm. Vegetation productivity is low in this area, but a drier type can be identified. This has a lower woody-biomass growing stock than the surrounding bushland and thicket vegetation, although annual productivity levels are similar.

With the exception of the bushland and thicket areas, the woody-biomass resources of Tanzania are considerable, although they show significant regional variations in their distribution. There is evidence of vegetation disturbance in all of the woodland and forest areas, and isolated patches of bushy thicket vegetation can be identified throughout the woodland and forest biomass classes.

Tanzania has been divided into seven biomass classes (see map, p. 142) using NOAA-7 GAC imagery and previous ecological studies and vegetation maps (Gillman, 1949; Moore, 1971; Polhill, 1968; Pratt *et al.*, 1966; Russell, 1962; and White, 1983). The classes are:

- Wet Miombo Woodland
- Wet Seasonal Miombo Woodland
- Dry Miombo Woodlands
- Cleared Miombo Woodland
- Coastal Forest Mosaic
- Semi-arid Steppe
- Semi-arid Dry Steppe

**DESCRIPTIONS OF INDIVIDUAL BIOMASS CLASSES**

**Wet Miombo Woodland**

Large areas of Tanzania are occupied by miombo woodland of various types. These have been sub-divided into three main categories: dry, wet seasonal, and wet miombo. These correspond to similar categories used elsewhere. The wet and dry variants differ largely in terms of the amount of photosynthetic activity taking place. Both show very marked seasonal variations and, consequently, there are implications regarding growing stock and annual levels of productivity. In the third type, Wet Seasonal

Miombo Woodland, the length of the season with little or no detectable photosynthetic activity is rather shorter than that for Dry Miombo Woodland and in some areas remains green throughout the year.

The largest areas of Wet Miombo Woodland are found on the border with Burundi, Rwanda and Zaire, extending from west of Lake Victoria to the Fipa Plateau. There are also large areas occurring within the Wet Seasonal Miombo Woodlands in southeastern Tanzania. Wet Miombo Woodland covers about 8.6% of the country or 71,941 km². It is not found in five districts (Arusha, Dodoma, Kilimanjaro, Mara and Singida) and is most extensive in Kigoma, Lindi, Rukwa and Ruvuma. There are also large areas in Morogoro and Tabora Districts. This biomass class also includes most of the closed forests found in the highland areas of Tanzania.

The NDVI values for Wet Miombo Woodland show little seasonal variation, ranging from about 175 to 185 in the wet season (highest in April) to a low point of about 160 in the dry season. The closed forests are now restricted to a few upland areas in the Rift Valley Escarpment, and the Southern and North-East Highlands. However, lowland forests do exist. They are mainly dry evergreen forests and moist forests are much less common. The forests can be divided into the following categories:

*Moist lowland forests* occur in some of the eastern highlands and also along Lake Tanzania. Moist montane forst is found above 1,200 masl and is restricted to areas with high rainfall. White (1983) classifies them as "Afromontane Rain Forests" as the effect of the seasonal drought stress is overcome by frequent dry-season mists. Consequently few trees, apart from secondary species, lose their leaves. One of the few documented "Afromontane Rain Forest" communities is the Western Usambara Mountains. Here the upper canopy varies in height from 25 to 45 m, a middle layer is found at 14–30 m, and the lowest dense canopy is found between 6 and 15 m. These tree layers are underlain by a shrub layer. The main tree species are *Aningeria adolfi-friedericii*, *Chrysophyllum gorungosanum*, *Parinari excelsa* and *Tabernaemontana johnstonii*.

The more extensive *dry evergreen forests* can also be divided into lowland and montane variants. Much of the Eastern Plateau and coastal regions were once covered in these forests, although there is debate as to whether it was an extensive natural forest cover.

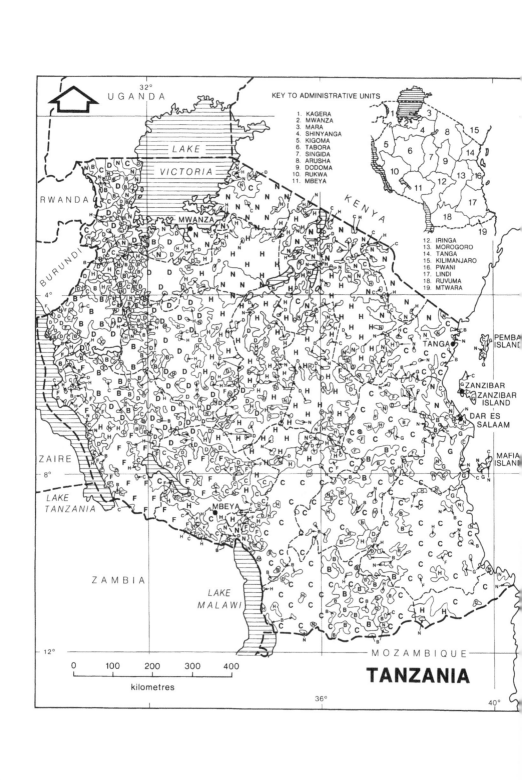

TANZANIA

However, due to over-exploitation it is now restricted to a few small remnants. These rarely exceed 25 m in height and the forest has a simple structure. It is floristically variable but two of the most commonly found species are *Isoberlinia giorgii* and *Parinari excelsa*.

*Dry montane evergreen forest* is restricted to the North-East and Central Highlands. A variety of mixed and pure-stand forest occurs in Tanzania. The most common is an undifferentiated dry montane forest with a canopy reaching about 20 m. Amongst the dominant trees are *Agauria* spp., *Ekebergia* spp., *Ilex mitis* and *Nuxia* spp. Transitional elements common to both lowland and upland dry evergreen forests are *Albizia gummifera, Brachylaena discolor, Manilkara obovata* and *Olea africana*. Pure-stand forests also occur and the main types are fire-induced *Arundinaria alpina* (Bamboo) thickets of 4–15 m high with associated woody species; the partially fire-resistant *Hagenia abyssinica* forest that is 9–15 m tall and is found between 1,800 and 3,400 masl; and the *Juniperus procera* forest in areas receiving between 1,000 and 1,500 mm rainfall per year, at altitudes of between 1,800 and 2,900 masl.

The forest areas occur mainly as small, isolated patches and their woody biomass potential can only be assessed in a local context. Both growing stocks and productivity are high but access is restricted by forest reservation and the fact that most are located in mountainous areas.

*Wet Miombo Woodland* occurs in areas with a mean annual rainfall in excess of 1,000 mm. In Tanzania, it is restricted mainly to the western parts of the Western Plateau but there are also large areas on the South-Eastern Plateau. Wet Miombo Woodland is usually above 15 m in height but rarely reaches more than 20 m. It is floristically richer than drier miombo woodland types. Evergreen species are also found more frequently. Where there are permanent streams, there may also be a strip of fringing riverine forest.

The main canopy species include *Brachystegia* spp. (which varies with site conditions and mean annual rainfall), *Commiphora apiculatum, Isoberlinia angolensis, I. paniculata* and *Terminalia mollis*. On the Eastern Plateau *I. tomentosa* is common, and on the South-Eastern Plateau *J. magnistipulata* is also generally found. The woodland is usually slightly open but a shrub layer rarely occurs. It is more often the case that ground cover is provided by *Andropogon* spp. and *Panicum maximum* grasses and saplings of the main tree species.

**Key to facing page:**

B Wet Miombo Woodland
C Wet Seasonal Miombo Woodland
D Dry Miombo Woodland
F Cleared Miombo Woodland
G Coastal Forest Mosaic
H Semi-Arid Steppe
N Semi-Arid Dry Steppe

Wet Miombo Woodland provides a readily accessible woody-biomass resource with high growing stock and high levels of productivity. However, extensive exploitation leads to floristic degeneration resulting in lower growing stocks and levels of productivity. This situation is already apparent in parts of Tanzania.

**Wet Seasonal Miombo Woodland**

Wet Seasonal Miombo Woodland accounts for most of the eastern third of Tanzania. It extends inland from the Kenyan border near Tanga until it reaches Lake Malawi and the Mozambique border. There are also outliers along the Kenyan border, in the dry central zone and in the south-west. The most significant outlier corresponds to the forests around the lower and middle slopes of Mount Kilimanjaro.

This is the largest biomass class in Tanzania covering almost a third of the country (32.7%), about 273,936 km$^2$. With the exception of Kagera it occurs in all districts, but is particularly significant in the Coast, Iringa, Lindi, Morogoro, Mtwara, Ruvuma and Tanga Districts. The greatest area (50,690 km$^2$) is found in Lindi District but there are also large tracts of this type of miombo woodland in Mbeya and Rukwa Districts.

Large parts of southern and eastern Tanzania have been placed in this biomass class whereas previous maps have assigned it to the Dry Miombo Woodland class (such as White, 1983). The evergreen nature of the woodland is more evident than in Dry Miombo Woodland, especially in hillier areas, and some parts seem to have virtually no period of the year that is without growth. Nevertheless, its productivity is affected by seasonality which differentiates it from the Wet Miombo Woodland; this is reflected in its phenology. Although NDVI values are high there is a marked seasonal dip in productivity between May and September, when NDVI values are as low as 150. During the wet season (November to April), they reach 180–185.

This class is floristically poorer than the Wet Miombo Woodland and is dominated by *Brachystegia spiciformis, B. boehmii, Julbernardia globiflora* and *J. magnistipulata*. The canopy rarely exceeds 15 m in height and is slightly more open than the Wet Miombo Woodland. In many places on the Coastal Lowlands and on the South-East Plateau, the miombo woodland has probably replaced dry evergreen forest. Trees from this older vegetation community can still be found, especially *Dalbergia* spp., *Ostryoderris* spp., *Pleurostylia* spp.,

*Sclerocarya* spp., and *Tamarindus* spp. In the extreme south-east of Tanzania, the miombo woodlands are replaced by structurally similar woodlands which are dominated by *Dalbergia* spp., *Lonchocarpus* spp. and *Millettia* spp.

Two of the forest zones around Mount Kilimanjaro fall into this class. On the lower slopes it corresponds to the lowland rain forest which is interspersed with banana and coffee plantations. At slightly higher elevations, on the middle slopes, the montane forests also fall into this class.

Woody-biomass resources of the Wet Seasonal Miombo Woodland are high, both in terms of growing stock and productivity, and physical accessibility is relatively easy. A danger of long-term ecological degradation exists if the woodland is over exploited.

**Dry Miombo Woodland**

Dry Miombo Woodland is restricted to the west of Tanzania. It extends south from Lake Victoria to just north of Lake Rukwa. It is bounded to the west by Wet Miombo Woodland and to the east by Semi-Arid Bushland and Thicket. This biomass class has about the same areal extent as Semi-Arid Dry Bushland and Thicket: 116,671 km² or 13.9% of the country. It covers over a quarter of the area in each of Kagera, Kigoma, Rukwa, Shinyanga, Singida and Tabora Districts, and there is also an extensive area in Mbeya. The largest occurrence is in Tabora District (34,802 km²).

The phenology of Dry Miombo Woodland is similar to that of Wet Seasonal Miombo Woodland. NDVI values vary from 175 to 180 in the November to April high-growth period and decline to 150 in August and September. The main diagnostic criteria for distinguishing the two classes is the lower integrated NDVI of the Dry Miombo Woodland which indicates generally lower growing stocks.

The dominant species of Dry Miombo Woodland, like those of the other types of miombo, are *Brachystegia* spp. (especially *B. spiciformis* and *B. boehmii*) and *Julbernardia globiflora*. Some tree species found in the drier biomass classes are also present, such as *Acacia brevispica, Acalypha fruticosa, Euclea divinorum, Grewia bicolor, Lantana camara, Ormocarpum trichocarpum* and *Terminalia mollis*. Generally however, the woodlands are floristically poor and less than 15 m high. In particularly adverse conditions they may be as low as 3 m. In the latter case, *Monotes* spp. and *Uapaca kirkiana* are often dominant.

Many of the miombo woodlands have been cultivated at some time in the past and this may be the reason for the relatively simple structure and floristic compositions. Locally however, considerable variation in miombo woodland occurs due to soil associations related to hydrological, soil, and topographic properties. Thus, well-developed miombo woodland is usually found on the well-drained interfluves and hill slopes (although with different dominant canopy trees). It is replaced, however, by *Combretum* wooded grassland on the lower slopes and grassland with no woody vegetation on the poorly drained *mbuga* soils of the swamps. On hardpans (soils with impermeable shallow subsoil horizons), *Acacia* and *Commiphora* dominate.

Dry Miombo Woodland is far more variable than either of the two wetter miombo variants. This leads to a greater variety in woody-biomass resource potential, the patterns of which are often complex and localized. In the best developed woodlands, growing stocks are high but there are limits imposed on exploitation by the strong seasonality in production patterns. In areas where the woodlands are under greater environmental stress, the growing stock and productivity levels will be significantly lower. Such areas include rocky hill slopes, lower slopes of hills and swamp margins. In such cases the reduction in woody-biomass resource potential is often related to a change in the ecological and florisitic composition of the woodland communities.

## Cleared Miombo Woodland

Cleared Miombo Woodland with significant amounts of shifting cultivation is restricted to south-west Tanzania. Extensive areas are found on the Fita Plateau, to the north of Lake Rukwa, and in the Kipenaere and Poroto Mountains. Cleared Miombo Woodland is found in 10 districts and covers 8.3% of the country, about 70,028 km². In most of the districts it provides only a small proportion of the land-cover but in Mbeya and Rukwa Districts, the area covered is significantly higher, 37.1% and 32.2% respectively. There are also large occurrences of this biomass class in Kigoma District.

It is phenologically distinct from the other types of miombo woodland. Wet-season productivity peaks in March when NDVI values reach 185; it is high from February to May. However, early wet-season productivity is low with NDVI values at as little as 150 in January. The usual dry-season decline in productivity is evident with NDVI values of 150–155 common between July and October.

This biomass class is dominated by a mosaic of woodland (at

various stages of regrowth) and cultivation plots. These plots are related to extensive shifting cultivation in which woodland is cleared and burnt prior to a cultivation cycle of 2–4 years. This mainly takes place in miombo or *Combretum* woodland. The regrowth of abandoned plots has been studied in adjacent areas of Zambia by Stromgaard (1985, 1986). Cultivation here was mainly of cassava, cowpea, maize, millet and groundnuts. After the first year of cultivation, three successional phases were identified:

i) *Phase 1* lasts up to one year after cultivation. The fields are still dominated by crops at this time but shrubs begin to invade after the first year. The main shrubs are *Euphorbia tirucalli* and *Smilax kraussiana*.

ii) *Phase 2* lasts from 2 to 6 years after clearance. Shrubs dominate the woody component, especially *Euphorbia tirucalli* and *Smilax kraussiana*. Other shrubs begin to invade and grasses become increasingly important, as groundnuts are harvested for up to three years. The grasses dominate the ground-cover by the sixth year, especially *Rhynchelytrum repens*.

iii) *Phase 3* lasts from 6 to 25 years. After 25 years, woody species have gained such a hold on the plots that a canopy woodland has formed. Shrubs are absent, and only grasses and sedges are found under the trees. The original woodland trees (*B. taxifolia*, *J. globiflora* and *Pterocarpus angolensis*) are absent. They are replaced by fire-resistant and fire-tolerant trees of the *Combretum* savannah. The dominant species were *B. spiciformis*, *Combretum mechowianum*, *Diplorhynchus condylocarpan* and *Syzygium guinensee*. The canopy these trees form enables *Uapaca* spp. to invade and, later in the succession (less than 25 years), the main components of the *Brachystegia-Julbernardia* woodland may reappear.

Overall, the picture is one of declining miombo woodland species and an invasion of savannah woodland species, especially those related to *Combretum* savannah. As pressure on the land increases, the recovery time for abandoned plots shortens and the proportion of savannah woodland to miombo woodland trees increases.

Fuelwood resources in this biomass class are generally moderate to high. In areas of woodland, both the growing stocks and productivity levels are high, but in the regrowth plots the stocks are more variable.

**Coastal Forest Mosaic**  The Coastal Forest Mosaic is most extensive along the coast between Dar-es-Salaam and Samanga. It is also found extensively in the three main islands of Mafia, Pemba and Zanzibar. This biomass class covers only 1.8% of Tanzania, about 15,102 km². It is restricted to the coast and is found as small areas in Dodoma, Lindi, Morogoro, Mtwara and Tanga Districts. A little over 60% of all of the Coastal Forest Mosaic occurs within Pwani District (9,269 km²). It is phenologically distinct from the other biomass classes, having high levels of productivity throughout the year (NDVI values of 175–185) and exhibiting little seasonality.

In the past, evergreen forests – both dry and moist – covered much of the lowland areas of Coastal Tanzania, the South-East and Makonde Plateaux, and the Eastern Usambara Mountains. Much of this has now reverted to miombo woodland but less disturbed evergreen forests can still be found. These are discussed in the Coastal Forest Mosaic class. Nevertheless, small patches do occur as outliers in the Wet and Wet Seasonal Miombo Woodlands.

Inland of Dar-es-Salaam, large areas of moist forest with transitional evergreen bushland and scrub forest are found extending southwards to the Matandu River. This is generally a floristically rich forest, 15–20 m tall, with emergents of 30–35 m high. The main canopy species are *Afzelia quanzensis* and *Erythrina sacleuxii*. The main emergents are *Albizia adianthiflolia, Elanites wilsoniana, Combretum schumannii, Julbernardia magnistipula, Lannea* spp. and *Manilkara sansibarensis*. As the coastal fringe of Mozambique is reached, the floristic diversity declines markedly although the forest structure, where it remains, is similar to that described above.

Along the coast between Kisiju and Kilwa Kivinjie there are extensive stands of mangroves, these are also found in Tanga District. The main species – *Avicennia* spp. and *Rhizophora* spp. – are present but distinct mangrove zones are developed along the East African coast. Other species, most notably *Bruguiera gymnorrhiza* and *Heritiera littoralis*, are also present.

The coastal forest mosaic contains high woody-biomass growing stocks and has high levels of productivity. Exploitation however is limited in the case of the mangroves by physical accessibility. The inland moist forest is a more fragile ecosystem than the mangroves and is easily prone to vegetation degradation if over-exploited. Evidence of this can be seen along the entire Tanzanian coastal zone as the coastal forest occurs as isolated relict areas.

**Semi-Arid Steppe**    The semi-arid steppe is an area of bushland and thicket which occurs as a large central triangular area stretching down from northern Tanzania almost to the Malawi border in the south. Significant outliers occur on the Mozambique border in the south and to the north-west of Lake Rukwa. It is the second most extensive biomass class in Tanzania covering 185,368 km² or 22.3% of the country. It occurs in all districts except Kigoma with large areas in Arusha (39,782 km²), Dodoma (26,876 km²), Shinyanga (21,156 km²) and Singida (29,861 km²). In these districts, and also in Iringa, Semi-Arid Steppe covers more than 25% of the land area.

Vegetation productivity is high from January to May (with NDVI values greater than 175) reflecting the actively growing vegetation. After May, productivity declines rapidly. Low rates of photo-synthetic activity are characteristic of the long dry season from July until October. The NDVI values recover slightly in November.

The vegetation is characteristically dense bushland, 3–7 m tall, with a few emergent trees of up to 20 m high. Evergreens usually form between 2.5% and 10% of the total trees and shrubs. Most woody plants have multiple stems and form bushes or small bush-like trees which are generally fire-resistant. Associated grasses reach 1.5 m in height and include tall *Hyparrhenia* species and shorter *Themeda*, *Setaria* and *Panicum* species. This vegetation is typical of the "Wooded Grassland", "Bushed Grassland" and "Bushland" classes of Moore (1971). The main tree species are *Acacia* (especially *A. gerrardi*, *A. hockii*, *A. mellifera*, *A. nilotica*, *A. seyal* and *A. tortilis*) and *Commiphora* (especially *C. africana*, *C. caerulea*, *C. molle* and *C. schimperi*). Other trees frequently found here include *Adansonia digitata*, *Boscia corimea*, *Cadaba farinosa*, *Cardia* spp., *Delonix elata*, *Lannea* spp., *Stericulia* spp. and *Terminalia* spp. A shrub layer is also found, the main species being *C. aculeatum*, *Grewia* spp. and *Maerua* spp. The bushland region also contains succulents and climbers.

Within the bushland areas, there are isolated patches of thicket vegetation ranging in size from small patches around old termitaria to large areas of several hundred square kilometres. Five main types can be identified, the most important being the *Itigi thicket* covering 620 km² of the Central Plateau. This is very dense, intertwined, deciduous thicket of 3–5 m in height, dominated by *Baphia burttii*, *B. massaiensis*, *Bussea massaiensis* and *Pseudoprosopis fischeri* with emergent evergreen and semi-evergreen trees of up to 8 m. Other thickets include:

i) *Cordyla* thicket in the east which is dominated by *Croton* spp., *Hipprocratea* spp., *Lannea* spp. and *Strychnos* spp.;

ii) *Commiphora* thicket;

iii) *Euphorbia* thicket; *and*

iv) a thicket on rocky hills dominated by *Dalbergia* spp., *Diospyros* spp., *Dombeya* spp., *Markhamia* spp., *Lannea* spp., *Strychnos* spp. and *Teclea* spp.

In drier upland areas, between the *Acacia-Commiphera* bushland and the montane woodlands and forest, an evergreen to semi-evergreen bushland is found. It is generally between 3 and 7 m in height and is dominated by *Ackanthera* spp., *Carissa edulis*, *Dodonaea viscosa*, *Euclea* spp., *Olea africana*, *Tarchonanthus camphoratus* and *Teclea* spp. Succulents are also commonly found.

Scrub forest is found locally in areas with slightly higher rainfall, for instance, communities are found around Lake Manyara and in the Western Usambara Mountains. The canopy is irregularly spaced but there is a dense under-storey usually reaching between 3 and 5 m in height. Dominant under-storey trees are *Commiphora* spp. with a variety of other species; the main emergents are *Adansonia digitata* and *Euphorbia* spp.

Edaphic wooded grassland found in valleys, floodplains and pans is also included in this class. These are predominantly grasslands with wooded areas, usually dominated by either *Acacia* spp. or palms (*Hyphaene* spp. and *Borassus* spp.). Along the larger rivers, some riparian forest of mixed floristic composition is found but it only accounts for a small area in total.

The potential for woody-biomass exploitation in this class is low. This is because although the actual growing stocks at the present time are relatively high for a semi-arid area, productivity levels are very low making it difficult, if not impossible, to reap an annual wood harvest. In addition, large areas of the steppe bushlands and thickets are reserved as game parks and many areas of thicket are almost impenetrable.

**Semi-Arid Dry Steppe**

Large areas of semi-arid dry bushland and thicket are found mainly in the northern part of Tanzania along the Kenyan border. A large tongue of this drier variant of the bushland and thicket communities also stretches into north-central Tanzania in Singida and Tabora Districts, and there is one large area on the coast north of Dar-es-Salaam. It is also found in smaller patches throughout Tanzania and therefore occurs in all districts. The class also includes the vegetation types on the upper slopes of Mount Kilimanjaro which have poor woody biomass stocks. However, its largest areal extent is in Arusha (30,689 km²), covering a little over a quarter of the district's area. It also covers high proportions of Kagera, Kilimanjaro, Mara, Mwanza and Tanga Districts. In addition, large areas are found in the Iringa, Pwani, Rukwa and Shinyanga Districts. In total, this class covers about 12.7% of Tanzania or 107,051 km².

Integrated NDVI values in the Semi-Arid Dry Steppe are considerably lower than those in the Semi-Arid Steppe, indicating much lower overall levels of photosynthetic activity. This class also exhibits a slightly different phenology. NDVI values decline from about 175 in January to lows of between 140 and 145 between July and October, after which they increase again. NDVI values are therefore lower than for the Semi-Arid Steppe phenological curves from February to May, although there is little difference in the length of the season of low productivity. This class has often been grouped with the previous category (e.g., White, 1983) but the lower integrated NDVI and dissimilar phenology seem to be clear diagnostic features of a different biomass class.

The vegetation consists of deciduous bushland and thicket with markedly seasonal vegetation growth and extensive grassy plains free from woody vegetation. This deciduous bushland is characterized in its natural state by stunted shrubs (2–6 m high) with only occasional emergent trees rising to a maximum of 10 m in height. It coincides with Moore's (1971) "Semi-Desert" category but also includes part of the area mapped as "Wooded Grasslands". The extension of this drier variant of the semi-arid woodland and thicket vegetation may be indicative of either long-term vegetation degradation or the adverse meteorological conditions in 1984, when the imagery used in the analysis was acquired. This areal extension is most noticeable along the Kenyan border in Mara and Kilimanjaro Districts.

The most common tree species are *Acacia* and *Commiphora* or saltbush (*Suaeda monoica*). Areas dominated by such vegetation have been described as "*Acacia-Commiphora* thorn savannah". However, detailed analyses of the variety of vegetation types occurring in the Serengeti Plains, which are contained within this vegetation class, are provided by White (1983) and point to the dominance of grassland communities. These are mainly responses to the soil conditions, specifically soils on young volcanic ashes, hardpan soils and black swelling soils (*mbuga*). Basal grass-cover in these areas varies from 15 to 45% and trees are rare, restricted to only *Acacia mellifera* bushes on young volcanic ashes. The grasslands are controlled by fire and if the interval between burning is greater than five years, *Acacia* can regenerate (Anderson and Talbot, 1957). After three to four years growth, they become fire-tolerant. Extensive secondary grassland communities therefore develop, which are essentially treeless and are controlled by burning, browsing and grazing.

The most extensive, wooded vegetation type is the "*Acacia-Commiphora* thorn savannah". It is dominated by *Acacia trees* (mainly *A. mellifera, A. seyal* and *A. tortilis*) and *Commiphora* spp. (mainly *C. madagascariensis* and *C. merkeri*). Important emergent trees include *Adansonia digitata* and species of *Euphorbia*. Scattered tussock grassland dominated by *Sporobolus robustus* is common under this wooded vegetation.

The dry woodland and thicket vegetation along the coast to the north of Dar-es-Salaam was mapped as "Wooded Grassland" by Moore (1971). Its inclusion in this biomass class is probably indicative of vegetation degradation since the early 1970s. The grassland component is related to the lower floodplains of the Mligasi, Msangasi and Wami Rivers. In the coastal regions, lowland forest and woodland has been commonly replaced by secondary scrubby woodland and grassland. The original *Diospyrus cornii* and *Manilkara mochisia* scrub forest that was 9–15 m tall has degraded to secondary deciduous bushland. The tree component is dominated by *Albizia anthelmintica, Acacia* spp., *Hyphaene compressa* and *Terminalia spinsoa*. This occurs as isolated areas all along the coastal region and also on the islands.

On the upper slopes of Mount Kilimanjaro, this class corresponds to the following altitudinal zones: the heather zone; the moorland zone; the alpine desert zone; and the summit zone. These are all characterized by low biomass reserves.

Biomass resources are restricted by low growing stocks and levels of productivity. Furthermore, woody plants in the *Acacia-Commiphora* bushlands are frequently spiny. This hinders movement and provides materials at times rather unsuitable for use as fuelwood. Considerable clearance for charcoal and fuelwood has taken place on the land covered by this biomass class, even though large areas are reserved as National Parks. Destruction of the vegetation by game, especially elephants, has occurred in recent years but this is a more serious problem in Kenya.

**BIOMASS SUPPLY**

Tanzania has high levels of woody-biomass resources, both in terms of growing stock and productivity. However, the sharp division of the country into three regions – two with high woody-biomass growing stocks and one with low growing stocks and an even lower productivity – polarizes the problems of woody-biomass supply. It also brings into focus the main area with low potential, central Tanzania.

This polarization of woody-biomass supply is clearly seen in the national summary statistics (see Tables 9.1 and 9.2). Over half of the growing stock and MAI is contained in the Wet Seasonal Miombo Woodland. This is restricted to the south-east and east of the country and only covers 32.6% of its land area. The supply problems are further distorted when all of the woodlands and forests are examined. Together, they account for under two-thirds of the land area (65.3%) but contain 91.2% of the woody-biomass reserves and 90.1% of the MAI. Their restriction to the south-east

**Table 9.1** TANZANIA: SUMMARY OF GROWING STOCK AND MAI DATA

| Biomass Class | Area | | Growing Stock | | MAI | |
|---|---|---|---|---|---|---|
| | (km²) | % | (mill t) | % | (mill t) | % |
| Wet Miombo Woodland | 71,941 | 8.6 | 512.7 | 15.3 | 15.2 | 14.2 |
| Wet Seasonal Miombo Woodland | 273,936 | 32.6 | 1,951.0 | 58.0 | 61.6 | 57.4 |
| Dry Miombo Woodland | 116,671 | 13.9 | 231.6 | 6.9 | 5.7 | 5.3 |
| Cleared Miombo Woodland | 70,028 | 8.3 | 234.9 | 7.0 | 7.1 | 6.6 |
| Coastal Forest Mosaic | 15,102 | 1.8 | 134.1 | 4.0 | 7.5 | 7.0 |
| Semi-Arid Steppe | 185,368 | 22.1 | 175.0 | 5.2 | 4.5 | 4.2 |
| Semi-Arid Dry Steppe | 107,051 | 12.7 | 121.8 | 3.6 | 5.8 | 5.4 |
| TOTAL | 840,097* | | 3,361.1 | | 107.4 | |

*Area excludes small farm cultivation and lakes.

153

and east (Wet Miombo Woodland, Wet Seasonal Miombo Woodland and Coastal Forest Mosaic) and the west (Wet Miombo Woodland, Dry Miombo Woodland and Cleared Miombo Woodland) essentially leaves central Tanzania (34.7% of the land area) with only 8.8% of the growing stock and 9.5% of the MAI. These resources are contained in the two steppe biomass classes.

Although areas of low growing stock and MAI are scattered throughout the woodlands, locally important extensive areas of low woody biomass potential occur in the following districts:

- Pwani (between the Miligasi River and Dar-es-Salaam)
- Mtwara (on the Macondes Plateau)
- Rukwa (Katavi National Park)
- Shinyanga (east of Kibondo)

Generally, however, the following districts have the greatest woody-biomass resource supply problems:

- Arusha
- Dodoma
- Iringa
- Kilimanjaro
- Mara
- Mbeya (south and east)
- Mwanza (east)
- Shinyanga (east)
- Singida

These areas of low biomass-resource supply include six of the major provincial towns in Tanzania, namely:

- Arusha
- Dodoma
- Moshi
- Mwanza
- Musoma
- Iringa

Conversely, a few districts hold enormous growing stocks – most notably Lindi, Morogoro, Rukwa and Ruvuma – and have equally high levels of producitvity (see Table 9.2).

The large reserved areas have a fundamental effect on the woody-biomass resource supply situation in Tanzania, mainly as National Parks and Game Reserves but also as part of the Forest

**Table 9.2** TANZANIA: SUMMARY OF GROWING STOCK
AND MAI DATA BY PROVINCE

| Region | Proportion of Growing Stock (%) | MAI (%) |
|---|---|---|
| Arusha | 4.4 | 4.7 |
| Dodoma | 2.5 | 2.8 |
| Iringa | 6.9 | 6.9 |
| Kagera | 2.8 | 2.7 |
| Kigoma | 4.3 | 3.3 |
| Kilimanjaro | 0.6 | 0.8 |
| Lindi | 11.7 | 11.9 |
| Mara | 1.0 | 1.3 |
| Mbeya | 7.5 | 7.2 |
| Morogoro | 14.0 | 14.2 |
| Mtwara | 3.9 | 4.3 |
| Mwanza | 1.3 | 1.3 |
| Pwani | 6.0 | 8.1 |
| Rukwa | 8.6 | 6.8 |
| Ruvuma | 11.5 | 11.5 |
| Shinyanga | 2.3 | 2.1 |
| Singida | 2.1 | 1.8 |
| Tabora | 4.4 | 3.9 |
| Tanga | 4.2 | 4.4 |

Estate. The National Parks and Game Reserves affect areas of both high and low woody-biomass potential but specifically restrict larger areas of low biomass resource in the low-potential areas outlined above, exacerbating an already critical situation. The major reserves and parks, and the main biomass classes affected, are outlined in Table 9.3.

Forest reserves are also important in Tanzania as they restrict access to miombo woodland and closed forests. However, this is really only a critical problem in the places with adjacent areas of low woody-biomass supply. Consequently, the vast areas of woodland under reservation in Morogoro, Lindi, Kigoma, Rukwa, Ruvuma, western Shinyanga and Tabora have much less effect on district woody-biomass supplies than the reservations in the steppe areas. Particularly important in the latter regions are the areally smaller

**Table 9.3**   GAME RESERVES AND NATIONAL PARKS IN TANZANIA

| Game Reserve(GR) or National Park(NP) | District(s) affected | Main biomass class(es) reserved |
|---|---|---|
| Biharamulo GR | Kagera | Wet and Dry Miombo |
| Katavi NP | Rukwa | Semi-Arid Steppe |
| Mikumi NP | Morogoro | Wet Seasonal Miombo |
| Lake Manyara NP | Arusha | Semi-Arid Steppe |
| Mkomazi GR | Kilimanjaro, Tanga | Semi-Arid Dry Steppe |
| Mt Kilimanjaro NP | Kilimanjaro | Semi-Arid Dry Steppe and Montane Forests |
| Ruaha NP | Iringa, Mbeya, Singida | Semi-Arid Steppe |
| Rungwa GR | Singida | Dry Miombo and Semi-Arid Steppe |
| Sadani GR | Pwani | Wet Seasonal Miombo |
| Selous GR | Pwani, Morogoro, Lindi | Wet and Wet Seasonal Miombo |
| Serengeti NP* | Arusha, Mara, Shinyanga | Semi-Arid and Semi-Arid Dry Steppe |
| Tarangire NP | Arusha | Semi-Arid Steppe |
| Ugalla River GR | Tabora | Wet and Dry Miombo |

*includes Ngorongoro Conservation Area and Musawa GR

reserves in Arusha, Dodoma, Iringa, Kilimanjaro and Tanga Districts. These are mainly closed forest and all are under severe local pressure because the surrounding regions have low woody and biomass resources.

The districts with potentially critical woody-biomass supply problems are (in tentative order of importance):

- Arusha
- Kilimanjaro
- Iringa
- Mara
- eastern Singida
- Dodoma
- south and east Mbeya
- eastern Shinyanga
- western Tanga

# 10.   Zambia

**GENERAL DESCRIPTION AND BIOMASS CLASSES**

Zambia has a complex mosaic of biomass classes. However, broad patterns within these classes can be determined from the interpretations of the satellite images. Much of the country is covered by miombo woodlands which are currently extending at the expense of other types of woodland. These range from closed canopy to open, sparse woodlands. Other types of woodland dominate the rest of the country and it is only around the swamps, on some of the alluvial soils, and in areas of extensive shifting cultivation, that the woodland component in the biomass classes is low.

Closed forest and dense woodlands account for 359,455 km² or 49.5% of the country. These forests and woodlands are mainly found in the north and west of the country on the Zambian Plateau and along the Malawian border. This area alone exceeds the closed forest and savannah woodland estimate for Zambia provided by Lanly (1981), who suggested the total area was 295,100 km² or 39.2% of the country. This suggests that the potential woody-biomass supply is greater than was previously thought. Furthermore, the woody-biomass potential is locally high in some of the more open-savannah woodland classes which in total account for 316,612 km² or about 42.6% of the country. The remaining 7.6% of the country is covered by two biomass classes dominated by grasses, herbs and low scrub vegetation. These are Scrub Woodland, and Swamp and Lake Vegetation. The open woodlands are found extensively in western Zambia (on the Kalahari Sands), in south-central Zambia (north of the Zambezi Escarpment overlooking Lake Kariba), in the Luangwa Valley, and in parts of the northern Zambia adjacent to Tanzania and Zaire. The small areas of grassland and low scrubby vegetation are restricted to the Zambezi Valley in western Zambia and around the main swamps.

The overall picture is one of high amounts of woody biomass in most areas with supply shortfalls evident only in parts of the following provinces:

- Central (including Lusaka)
- Copperbelt
- Northern
- Southern
- Western

Most of these areas have low population densities and few large towns. If the population densities and urban concentrations are taken into account, the following provinces and towns appear to have shortfalls in woody-biomass supply:

- eastern Central and Lusaka
- eastern Copperbelt
- central Northern
- Southern
- central Western

The main cities and towns occurring in low woody-biomass supply areas are:

- Kabwe, Kafue and Lusaka (Central and Lusaka)
- Kitwe, Luanshya and Ndola (Copperbelt)
- Kasama and Mbala (Northern)
- Choma, Kalomo, Livingstone, Nazabuka and Mouze (Southern)
- Mongu (Western)

The country has been divided into nine biomass classes (see map on p. 160), using NOAA-7 AVHRR GAC data. There have been extensive and detailed studies of the vegetation of Zambia undertaken by D.B. Fanshawe between 1967 and 1972 and covering each district. This work was summarized by Fanshawe (1969) and forms the basis of a set of nine 1 : 100,000 vegetation maps compiled by A.C.R. Edmonds in 1976. These sources were used extensively but the work of Trapnell (1953), Trapnell and Clothier (1957), Trapnell et al. (1952), and White (1965) were also referred to. The biomass classes are:

- Wet Miombo Woodland
- Seasonal Miombo Woodland
- Dry Miombo and Munga Woodland

- Degraded Miombo Woodland
- Dry Evergreen Forest
- Kalahari Woodland
- Mopane Woodland
- Scrub Woodland
- Swamp and Lake Vegetation

**DESCRIPTIONS OF INDIVIDUAL BIOMASS CLASSES**

**Wet Miombo Woodland**

The Wet Miombo Woodland class is distributed widely throughout Zambia, being found in all provinces. It is especially important in Central, Eastern, Luapula, Northern and North-Western Provinces, accounting for over 30% of the vegetation in each case. It is the largest biomass classes in the country, covering 223,942 km², or 30.9% of the total land area.

It is possible to differentiate four biomass classes in which miombo woodland is the dominant or co-dominant vegetation type. This is done on the basis of integrated NDVI values and the seasonal variations in NDVI. In approximate order of growing stock and productivity these are Wet Miombo, Seasonal Miombo, Cleared Miombo, and Dry Miombo. The latter two are included in biomass classes with other vegetation types. This division does not fit the floristic division of Trapnell (1947, 1957) or Lees (1963) as it is based on biomass parameters rather than floristic composition.

Wet Miombo Woodland has high levels of productivity throughout the year ranging from NDVI values of 175–185 between November and April, and lower values in the dry season, going down to about 160 in September.

Wet Miombo Woodland is generally a two-storey closed semi-evergreen woodland. The main upper-storey dominants are *Brachystegia* spp., *Isoberlinia* spp., *Julbernardia* spp. and *Marquesia macroura*. Frequent associates are *Anisophyllea pomifera*, *Erythrophleum africanum*, *Parinari curatellifolia* and *Pterocarpus angolensis*. The lower storey is structurally less well-defined but is floristically more diverse. Underneath this can be found either a 0.6–1.3 m grass and suffrutex layer or a dense evergreen thicket reaching heights of 3.5 m. A variety of suffrutices are common in miombo woodlands and the grass-cover varies both in density and height according to season.

The occurrence of dense, evergreen, thicket under-storey is indicative of the fact that miombo woodlands have replaced dry

159

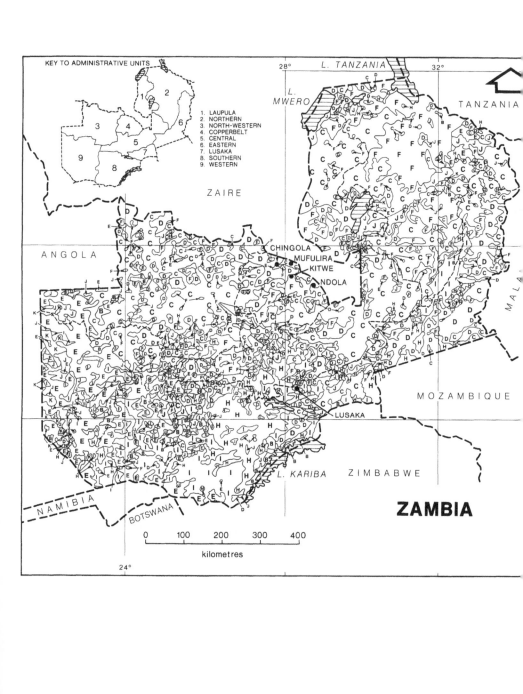

KEY TO ADMINISTRATIVE UNITS

1. LAUPULA
2. NORTHERN
3. NORTH-WESTERN
4. COPPERBELT
5. CENTRAL
6. EASTERN
7. LUSAKA
8. SOUTHERN
9. WESTERN

L. TANZANIA

L. MWERO

TANZANIA

ZAIRE

ANGOLA

CHINGOLA
MUFULIRA
KITWE
NDOLA

MALA

MOZAMBIQUE

LUSAKA

L. KARIBA

ZIMBABWE

NAMIBIA

BOTSWANA

ZAMBIA

28°          32°

24°

| 0 | 100 | 200 | 300 | 400 |

kilometres

evergreen forest in many parts of Zambia. Fanshawe (1969) concludes that miombo is an invasive woodland type replacing many types of the dry evergreen forests that covered Zambia in wetter periods. The main woody-thicket species are *Canthium burttii*, *Cassipourea* spp. and *Chrysophullum magalismontanum*, as well as many shrubs. Often, the only dominant miombo woodland overwood tree is *B. spiciformis*.

Two main variants of Wet Miombo Woodland are apparent in Zambia. On the deeper soils of the Zambian plateau, the main canopy dominates are *B. boehmii, B. floribunda, B. spiciformis, B. utilis, I. angolensis* and *J. paniculata*. On shallower soils on hills and escarpments and in extensive pockets of sand (*isengas*), the canopy dominants change. At first to *B. glaucescens* (in the south) or *B. microphylla* (in the north) and finally to *B. taxifolia* and *Cryptoseplum exfoliatum*. This change is even more marked in the shrub and grass components. The "hill miombo" is found on the Muchinga Escarpment and the Bwingmfumu Hills.

The woody-biomass reserves of the Wet Miombo Woodland are very high, both in terms of growing stock and levels of productivity. It is an aggressive vegetation type that can withstand quite large-scale exploitation.

**Seasonal Miombo Woodland**

Seasonal Miombo Woodland is closely related to the Wet Miombo Woodland, the main difference being its marked seasonality. This class occurs on the Zambian Plateau, the Zambezi Escarpment, and extensively along the Mozambique and Malawi borders. It is the second largest biomass class in Zambia, accounting for 125,715 km$^2$ or 17.3% of the country. It occurs in all provinces but is areally most extensive in Eastern and North-Western Provinces.

The main diagnostic criteria for seasonal miombo woodland is a marked dry-season decline in photosynthetic activity. High NDVI values (greater than 180) are typical of the period from November to April, but after this there is a long die-back period. Low values of NDVI (and therefore low rates of photosynthetic activity) are found in August and September.

The woodland is similar in structure to the Wet Miombo Woodland but the canopy is more open and there is a far greater proportion of deciduous trees. Consequently, species such as *B. allenii, B. bussei, Burkea africana, I. angolensis, J. globiflora* and *Terminalia sericea* appear with more frequency in the canopy. The shrub and grass layers may be better developed and are floristically different. The areas of dry evergreen thicket do not occur in this more open miombo woodland because of the longer dry-season moisture-stress period.

Extensive areas of "hill miombo" woodland occur in this biomass class and are found particularly on the hills along the border of Zaire, the hills along the Mozambique border in North-Western Province and the Zambian extension of the Nyika Plateau. In these areas, the trees are under severe drought-stress due to the thin, stony soils. The canopy is dominated by *B. glaucescens, B. microphylla, B. taxifolia* and *Cryptosepalum exfoliatum.*

Seasonal Miombo Woodland is almost as resilient as Wet Miombo Woodland. The growing stock is high, as are the productivity levels, although the latter is marred by a dry season fall-off in production which is a response to moisture stress.

**Dry Miombo and Munga Woodland**

The Dry Miombo and Munga Woodland biomass class is found in all districts except Copperbelt, Luapula and North-Western. The greatest extent is to the south-west of Lusaka, and to the north of Lake Kariba in Lusaka and Southern Provinces. There are also locally important isolated areas in the Eastern and Western Provinces. It only covers 7.3% of Zambia, an area of about 53,085 km².

These woodlands show a marked dry season fall-off in productivity and a lower wet-season rate. Consequently, the growing stock and productivity is lower than either the Wet or Seasonal Miombo Woodlands. Productivity levels are high from December until May, with NDVI values of around 175. This is followed by a rapid decline in June and July, resulting in low activity rates from July until October.

The Dry Miombo Woodland component of this biomass class has both natural and disturbed elements. It is commonly found on alluvial sands and flats, along the Zambezi Valley, on the Kafue Flats, and on the Kalahari Sands. In the first two areas, it is closely associated with Munga Woodland. The canopy in such woodlands is dominated by *Erythrophleum africanaum,* usually combined with *B. allenii, B. bussei, Burkea africana, I. angolensis, J. globiflora* and *Terminalia sericea.* The woodland has a relatively open canopy of widely spaced deciduous trees. This enables a diverse under-storey to develop, of which the common trees are *Baphia massaiensis, Combretum elaeagnoides,*

*Crossopteryx febrifuga, Dalbergiella nyassae, Diospyros kirkii, Diplorhynchus condylocarpon* and *Pseudolachnostylis macrouneifolia*.

Munga Woodland is equally widespread within this biomass class. It is particularly common on the Kafue Flats, in the Southern District, and along the Lunsemjwa, lower Luangera and Zambezi valleys. It is a type of savannah woodland with an open, park-like appearance. There are either one or two woody layers, both deciduous, with emergents reaching 18 m in height. Particularly important in the canopy are *Acacia, Combretum* and *Terminalia* trees, although it can be floristically quite diverse. Sometimes, a dense woody under-storey reaching 4.5 m is found. The bushes here are deciduous or semi-deciduous and, once again, although floristically diverse this under-storey is dominated by *Acacia* spp. and *Combretum* spp. Occasionally climbers occur but there is always a tall, thick grass layer which is burnt annually in most areas.

In some Munga Woodland areas there are woody thickets. These are related to soil and water conditions and usually include the following trees: *Commiphora mollis, Euphorbia candelabrum, Markhamia obtusifolia* and *Schrebera trichoclada*.

Munga Woodland shows some regional and local variations in Zambia (Fanshawe, 1969) but it can be regarded generally as an invasive secondary woodland. It appears to have developed by the invasion of woodland trees onto alluvial grasslands and all tree species are fire-resistant. Many other woodland types are being invaded by Munga Woodland, especially riparian woodland.

Both Dry Miombo and Munga Woodland have low growing stocks and levels of productivity, mainly because of a strong seasonal drought. Both woodland types are expanding at the present. The more aggressive Munga Woodland increases the woody component of alluvial grassland but in other cases replaces more closed woodland with a wooded grassland. Exploitation is limited by low levels of productivity.

**Degraded Miombo Woodland**

Extensive areas of Degraded Miombo Woodland, and related woodland and wooded grassland vegetation types, are found to the north of the Kafue Flats, in the Copperbelt, and in northern Zambia adjacent to the Tanzanian border. It is most areally extensive in Central, Copperbelt, Luapula and Northern Provinces. In total, this class covers 110,161 km² or about 15.2% of the country.

This biomass class is phenologically distinct from the other

Zambian woodlands. Productivity levels are generally high from February to May, and peak in March with NDVI values of 185. A strong dry-season decline is evident and NDVI values fall to between 150 and 155 from July to October.

In these areas, woodland is destroyed by a type of shifting agriculture practised by the Bemba. It is known as *chitemene* and occurs mainly in miombo and *Combretum* woodlands. The regrowth of abandoned plots has been studied by Stromgaard (1985a, 1986). Cultivation is mainly based on cassava, cowpea, maize and millet, with groundnuts planted in the second year. Three successional phases were identified:

   i)   *Phase 1*, lasting up to one year after cultivation. The fields are still dominated by crops but shrubs begin to intrude after the first year. These are mainly *Euphorbia tirucalli* and *Smilax kraussiana*.

   ii)   *Phase 2*, lasting from 2 to 6 years after clearance. Shrubs dominate the woody component, especially *Euphorbia tirucalli* and *Smilax kraussina*. Other shrubs begin to invade and grasses become increasingly important as crops are harvested for up to three years. The grasses dominate the ground cover by the sixth year, especially *Rhynchelytrum repens*.

   iii)   *Phase 3*, lasting from 6 to 25 years. After 25 years, woody species have gained such a hold on the plots that a canopy woodland has formed. Shrubs are absent, and only grasses and sedges are found under the trees. The original woodlands dominants (*B. taxifolia, J. globiflora* and *Pterocarpus angolensis*) are replaced by fire-resistant and fire-tolerant trees characteristic of the *Combretum* savannah (e.g., *B. spiciformis, Combretum mechowianum, Diplorhynchus condylocarpan* and *Syzygium guineense*). The canopy these form enables *Uapaca* spp. to invade and, later in the succession (more than 25 years), the main components of the *Brachystegia-Julbernardia* woodland may reappear.

The overall picture is one of declining miombo woodland species and an invasion of savannah woodland species, especially those related to *Combretum* savannah. As population pressure increases, the recovery time for abandoned plots shortens and the proportion of savannah woodland to miombo woodland grows.

Other vegetation types are commonly found in the Degraded

Miombo Woodland biomass class. In northern Zambia these include *Itigi* Forest, Lake Basin *Chipya* Woodland, and Grassland and Terminitary Vegetation associated with highland areas.

*Itigi* forest is found along the Zaire border, north of the Meru Swamp. It is composed of 6–12 metre-high deciduous trees, mainly *Baphia massaiensis, Boscia angustifolia, Burttia prunoides, Bussea massaiensis, Diospyros mweroensis,* and succulents. Underneath is a deciduous to evergreen thicket, 3 to 4.5 m high, dominated by *Boscia mossambicensis* and *Teclea fischeri.* It is a pre-climax forest, unable to attain dry deciduous forest forms because of poor drainage conditions.

Disturbance of *Itigi* Forest results in the formation of Lake Basin *Chipya* Woodland and Scrubland. This is a three-tier woodland with an open deciduous canopy which can reach heights of 27 m. The main canopy dominants are *Albizia antunesiana, Burkea africana, Combretum collinum, Erythrophleum africanum, Parinari curatellifolia, Pterocarpus angolensis* and *Terminalia sericea.* This is underlain by a floristically diverse, evergreen to semi-deciduous under-storey of 6–12 m in height. The main dominants at this level are *Combretum* spp., *Diplorhynchus condylocarpon, Markhamia obtusifolia, Piliostigma thonningii, Pseudolachnostylis maprouneifolia* and *Syzygium guineense.* Under the second storey are a 2–3 m high shrub layer of 2–3 m high and a rich ground flora that reaches 0.6–2 m. It occurs extensively in Luapula Province along the Zaire border and around the Bengwelu Swamp.

The grasslands in the Northern Province have been extensively studied by Vesey-Fitzgerald (1963) who identified a variety of ecological sites. These are mainly pure stand grassland with no tree-cover. The only woody biomass in these large grassland areas in Northern Province is associated with old termite mounds.

With the exception of the grasslands, this biomass class has high woody-biomass growing stocks and moderate rates of productivity. However, most of the vegetation types rapidly degenerate under disturbed conditions sometimes, but not always, to the detriment of growing stocks and productivity.

**Dry Evergreen Forest**  Dry Evergreen Forests once covered large areas of Zambia before miombo woodlands invaded as the climate became drier. They are now mostly restricted to western Zambia with the largest areas being found in North-Western and Southern Provinces. They only account for 1.3% of the country, covering 9,798 km².

They show little variation in productivity throughout the year. NDVI values never go below 170 although a slight dry-season depression, from wet-season NDVI values of about 180, is typical of the period from May until November.

Well-developed forests have three tiers: a closed evergreen canopy 25–27 m tall, with emergents; a discontinous evergreen under-storey reaching 9–15 m; and a dense evergreen shrub-scrambler thicket of 1.5–6 m in height. Three geographical locations for dry evergreen forest have been identified in Zambia (Fanshawe, 1969) – the Zambian Plateau, the Kalahari Sands and the Lake Bengwelu Basin. The latter zone is dominated by Lake Basin *Chipya* which was considered in the previous biomass class. The Zambian Plateau forests are now quite small in areal extent and the main evergreen forests are now found at the boundary of the Kalahari Sands and Zambian Plateau. The main environmental factor contributing to the evergreen nature of the forests is their ability to retain sufficient soil-moisture reserves in the dry season, either by high storage levels or high total rainfall.

Three types of dry evergreen forest have been identified (Fanshawe, 1969):

i) *Parinari* forest is found on the Zambian Plateau and has two main canopy trees, *Parinari excelsa* and *Syzygium guineense*. Common under-storey trees are *Aidia micrantha*, *Chrysophyllum magalismontanum*, *Olea capensis*, *Tabernaemontana angolensis* and *Teclea nobilis*.

ii) *Marquesia* forest is found in the Lake Bengweulu Basin. The canopy dominants are *Anisophyllea pomifera*, *Marquesia* spp., *Podocarpus milanjianus* and *S. guineense*. The main under-storey trees are *A. micrantha*, *Combretum celastroides* and *Olea capensis*.

iii) *Cryptosepalum* forest canopy dominants are restricted to *C. pseudotaxus* and *Guibourtia coleosperma* but other trees, mainly *Marquesia* spp., *P. excelsa* and *S. guineense* are found. The main under-storey representatives are *Baphia massaiensis*, *Chrysophyllum magalismontanum*, *Diospyros undabunda* and *O. capensis*.

All of the dry, evergreen forests show signs of invasion by miombo-woodland trees at their edges. This suggests a high level of ecological fragility under present climatic conditions. Disturbance by anthropogenic causes leads to the formation of *chipya* woodland, each type of evergreen forest creating a slightly different type.

The evergreen forests of Zambia have very high growing stocks and annual levels of productivity. However, as they are in an ecologically marginal situation any exploitation will quickly lead to the formation of either *chipya* or miombo woodland.

**Kalahari Woodland**

Kalahari Woodland is restricted to western Zambia and is mainly found in North-Western, Western and Southern Provinces. Other occurrences of this biomass class are probably representative of *chipya* woodland. In total, the class covers 79,212 km², about 10.9% of the country.

The phenology of Kalahari Woodland shows a marked seasonality. Wet season NDVI values of about 175 are common from November to April, after which they gradually decline until rates of about 150 are reached in August and September.

Two types of Kalahari Woodlands have been identified by Fanshawe (1969). Other types of Kalahari Sand vegetation are dealt with in the Scrub Woodland class. *Guibourtia* woodland is a two-storey open woodland with deciduous to semi-deciduous, floristically-rich upper canopy of 18–24 m tall which includes *G. coleosperma*, a variety of deciduous trees and invasive miombo-woodland species. The under-storey is composed of a thicket, 1.3–2.6 m high, of small trees and shrubs. Climbers and scramblers are scarce and the grasses vary in density.

*Burkea-Erythrophleum* woodland has a more open canopy than *Guibourtia* woodland and the under-storey is unstratified. It is floristically less diverse and dominated by *Burkea africana* and *Erythrophleum africanum*. Both these trees are also found in *Guibourtia* woodland and *G. coleosperna* is common in this type of woodland. Other trees common to both types of woodland are *Amblygonocarpus andongensis*, *Combretum mechowianum*, *Cryptosepalum exfoliatum*, *Dialium engleranum* and *Monotes* spp. The miombo-woodland elements are rarely found in *Burkea-Erythophleum* woodland.

Other types of Kalahari woodland are mainly *Burkea-Diplorhynchus* and *Diplorhynchus* scrub. The former, the most woody of these, is dealt with in this biomass class. In *Burkea-Diplorhynchus* scrub there is only open woody scrub vegetation, 4.5–6 m high, dominated by bushy *Burkea africana* and *Diplorhynchus condylocarpan*.

These woodlands are all related, each securing a slightly different ecological situation. In general, the deciduous Kalahari Woodlands have developed because of the destruction of the natural *Baikiaea* forests which are now restricted to extreme south-west Zambia.

Where *Baikiaea* forest remains, it has two storeys and an open or closed deciduous canopy reaching 9–18 m. The two main canopy trees are *B.plurijuga* and *Pterocarpus antunesii*, but invasive *Acacia* and *Combretum* species are common. It is underlain by a shrub layer – the *mutenwa* – a deciduous thicket reaching over 1 m in height. This contains *Acacia ataxacantha, Baphia massaiensis, Barhinia macranha, Combretum* spp., *Dalbergia martinii* and *Popowia obovata*. There is a dense wet-season ground flora.

The *Baikiaea* forest area has declined rapidly in recent years due to cultivation, burning, competition from *Cryptosepalum* forest and, most importantly, timber extraction. *B. plurijuga* is a valuable timber tree – the "Zambezi Teak". The destruction of the forest leads to *chipya* and other types of Kalahari Woodland, or a secondary *Baikiaea* woodland with elements from adjacent Munga and Kalahari Woodlands.

All of these types of deciduous woodlands have moderately high growing stocks and relatively low levels of productivity. Consequently, only low rates of exploitation can be carried out to avoid widespread vegetation destruction.

**Mopane Woodland**

Mopane Woodland is widespread in Zambia, covering 69,000 km² or 9.5% of the country. Major areas are found in the Luangwa Valley and in southern Zambia to the west of Lake Kariba; small patches occur in other areas. It is most important in Southern and Western Provinces.

It has a distinctive phenology. NDVI values decline from a high of 175 around the turn of the year, to a low of about 150 in September and October. The decline in productivity is gradual and relatively constant.

Mopane is a single storey woodland of 6–18 m tall, with an open deciduous nature. The dominant tree is *Colophospermum mopane*. Although it often occurs as pure stands, occasionally Munga Woodland trees invade, especially *Acacia nigrescens, Adansonia digitata, Combretum imberbe, Kirkia acuminata* and *Lannea stuhlmannii*. Lower trees and shrubs are usually absent, and grass and herb cover is dominant only locally. It is restricted to alkaline soils on valley floors which flood in the wet season and dry out in the dry season. Destruction of

Mopane Woodland leads to a drastic reduction in the woody-biomass component as it is replaced by coarse, tussock grassland.

Both the growing stock and productivity of Mopane Woodland are limited. Combined with a drastic reduction in the woody biomass when overexploited, this suggests that exploitation for firewood should be restricted.

**Scrub Woodland**

Scrub Woodland is restricted to the tributaries of the Zambezi on the Angolan border and has equivalents in the *Chanas da borracha* grasslands in Angola. The main areas covered by this class are along the Luachi, Luangingo, Luanbimba, North and South Lueti, and Lungewebumu Rivers. The largest areas are therefore found in Western Province. Elsewhere, it probably represents small isolated areas of scrubby thickets. In total it covers 9,800 km² or about 1.3% of the country.

Productivity levels are low throughout the year, and the wet and dry seasons are barely distinguishable. NDVI values in the wet season vary from 150 to 160, whilst dry-season values rarely exceed 150.

Three elements related to the Scrub Woodland class have been identified by Fanshawe (1969):

i) The *Burkea-Diplorhynchus* Scrub is a 4.5–6 metre-high, open woody scrubland dominated by *B.africana* and *D.condylocarpon*. It occurs on the highest ground in the river valleys.

ii) At slightly lower elevations, *Diplorhynchus* Scrub is found. This reaches 2 m in height and is very open with about 12–25 scrubby trees per hectare. There is a thick ground-cover of suffrutices. The trees are mainly *D. condylocarpon* and *Hymenocardia acida*.

iii) The final category is *Parinari* suffrutex savannah. This is a ground carpet of suffrutices reaching 30 cm in height with no emergent trees.

In other areas, and along the river valleys, the biomass class represents grassland and, to a lesser extent, scrubby thicket. The grasslands are generally floristically diverse but are without woody species. They are usually controlled either by flooding, poor soil or burning.

This biomass class has very low growing stocks and low productivity levels; fuelwood resources are minimal.

## Swamp and Lake Vegetation

Swamp and Lake Vegetation is found along the shores of the main lakes –Bengweulu, Kariba, Mweru and Tanzania – and in the large swamps typical of the Zambian Plateau – the Bengweulu, Lukanga, and Mweru Swamps being the most important. It is found in all provinces but is particularly important in Northern, North-Western, Southern and Western Provinces. In all it covers 6.3% of the country, or 46,140 km².

Productivity levels show a slight seasonality with high NDVI values ranging from about 160 to 170, which occur between February and May. Productivity declines after this, reaching a low point in September.

In many areas, the swamp and lakeside vegetation is a grassland with few or no woody plants. However, some areas of swamp forest do occur. These are mainly three-storey closed evergreen forests with a canopy that reaches 27 m and is dominated by *Ilex mitis, Syzygium* spp. and *Xylopia* spp. It is underlain by a discontinuous evergreen under-storey, 9–18 m high, and a dense evergreen shrub layer which reaches 4.5 m. The forest floor is either bare or covered by herb stands.

All swamp forests are controlled by high groundwater levels. They are also small in extent, varying from 1 to 120 ha. Fanshawe (1969) estimated that only 380 km² of swamp forest exist in Zambia. Consequently, although they have high growing stock and productivity levels, they are only locally important as woody-biomass resources. Generally however, this biomass class is typified by very low woody growing stocks and levels of productivity.

## BIOMASS SUPPLY

Zambia possesses substantial woody-biomass resources both in terms of existing growing stocks (2744.68 million tonnes) and MAI (83.3 million tonnes/year) as can be seen in Table 10.1. The national summary statistics give an areal distribution of 3,776 tonnes/km² and a per capita distribution of 394 tonnes per person (based on 1986 population estimates).

As with most other SADCC countries, there is an uneven distribution of woody-biomass resources by both class and province. Approximately a third of Zambia (30.9%) is covered Wet Miombo Woodland. This occurs mainly in the north and west of the country, and the area contains almost two-thirds of Zambia's growing stock (58.1%) and MAI (60.5%) as can be seen in Table 10.1.

**Table 10.1**  ZAMBIA: SUMMARY OF GROWING STOCK AND MAI DATA

| Biomass Class | Area (km²) | Area (%) | Growing Stock (mill t) | Growing Stock (%) | MAI (mill t) | MAI (%) |
|---|---|---|---|---|---|---|
| Wet Miombo Woodland | 223,942 | 30.9 | 1,594.6 | 58.1 | 50.4 | 60.5 |
| Dry Miombo and Munga Woodland | 53,085 | 7.3 | 50.1 | 1.8 | 1.3 | 1.6 |
| Seasonal Miombo Woodland | 125,715 | 17.3 | 249.5 | 9.1 | 6.1 | 7.3 |
| Dry Evergreen Woodland | 9,798 | 1.3 | 69.8 | 2.5 | 2.2 | 2.6 |
| Degraded Miombo Woodland | 110,161 | 15.2 | 369.6 | 13.5 | 11.1 | 13.3 |
| Mopane Woodland | 69,000 | 9.5 | 255.1 | 9.3 | 7.5 | 9.0 |
| Scrub Woodland | 9,800 | 1.3 | 22.9 | 0.8 | 0.8 | 1.0 |
| Swamp and Lake Vegetation | 46,140 | 6.3 | 0 | 0 | 0 | 0 |
| Kalahari Woodland | 79,212 | 10.9 | 133.0 | 4.8 | 3.9 | 4.7 |
| TOTAL: | 726,853 | | 2,744.6 | | 83.3 | |

The other woodlands fall into two groups: those with almost equal proportions of area, growing stock, and MAI; and those with growing stocks and MAIs that are lower than the areal proportions. The first of these two categories contains the Kalahari and Mopane Woodlands as well as the Dry Evergreen Forests. Together they cover 158,009 km² (21.7% of Zambia) and contain a growing stock of 457.9 million tonnes (16.6%) and an MAI of 13.6 million tonnes/ year (16.3%). However, Seasonal Miombo, Dry Miombo and Munga, and Degraded Miombo Woodland, cover 39.8% of Zambia but only contain 24.4% of the national growing stock and 22.2% of the MAI. It is these latter woodlands that are responsible for the potential shortfalls in woody-biomass supply in the more highly populated areas of the country.

Two biomass classes – Scrub Woodland, and Swamp and Lake Vegetation – are grossly under-represented in the woody-biomass supply statistics. Although they account for 7.6% of the country, they only contain 0.8% and 1% of the growing stock and MAI respectively.

The inequitable woody-biomass resource distribution is compounded by forest reserves and national parks. Fortunately, many of the national parks are in sparsely populated areas. These include Kasanka, Liuwa Plain, Luambe, North and South Luangwa, Lukusuzi, Sioma Ngweze, and West Lunga National Parks.

However, seven occur in or adjacent to areas of high fuelwood demand. In areas of naturally low biomass resources this undoubtedly causes supply problems. They are the Blue Lagoon, Lochinvar and Kafue National Parks in Central and Southern Provinces; and Isangano, Lusenga Plain, Mweru Wantipa and Sumba National

**Table 10.2** ZAMBIA: GROWING STOCK AND MAI DISTRIBUTION BY PROVINCE

| District | Proportion of Growing Stock (%) | MAI (%) |
|---|---|---|
| Central | 12.0 | 12.1 |
| Copperbelt | 5.1 | 5.1 |
| Eastern | 8.0 | 7.7 |
| Luapula | 7.9 | 8.0 |
| Lusaka | 4.2 | 4.0 |
| Northern | 23.4 | 23.8 |
| North-Western | 19.9 | 20.1 |
| Southern | 7.2 | 6.9 |
| Western | 12.2 | 12.2 |

Parks. The Zambia forest estate is extensive and includes National and Local Forests, both of which cover Protected and Reserved Forests. In 1975, there were 381 individual areas of protected or reserved forest (Government of Zambia, 1975). The majority of the large reserves and protected areas are in the Wet Miombo Woodlands and Dry Evergreen Forests. There are notable exceptions, however, in the Kalahari, Mopane and Seasonal Miombo Woodland Classes. When matching woody-biomass supply to high demand areas at the scale of this survey, it is vitally important to consider which provinces have high areal proportions of protected and reserved forests. Undoubtedly, there are serious problems in Copperbelt Province where most of the better woodland is part of the Forest Estate. Furthermore, there are localized problems in the following areas:

- around Livingstone
- along the line-of-rail from Kafue to Zimba (including Choma)

- around Mbala
- around Chipata

The final point to consider in the assessment of woody-biomass supply are wood preferences for firewood and charcoal, and conflicting end-uses. Fanshawe (1962) has examined in detail the 50 most common Zambian trees. Seven of these were considered important fuelwood and charcoal resources but all have conflicting end-uses (see Table 10.3).

**Table 10.3** END-USE OF PREFERRED FIREWOOD AND CHARCOAL TREES IN ZAMBIA

| Tree | Main vegetation types found in: | Other end-uses |
| --- | --- | --- |
| Balanites aegyptica | Mopane | tool handles, poles, cooking oil and soap |
| Bridelia micrantha* | Drier forests, near rivers | timber, poles, general joinery, boats, dyes and traditional medicines |
| Dichrostachys cinerea* | Munga, Baikiaea and drier woodlands | fencing, tool handles, cattle food baskets and traditional medicines |
| Diplorrhynchus condylocarpon | Miombo and degraded Kalahari Woodland | furniture, tool handles, poles, gum and traditional medicines |
| Piliostigma thonningii | All woodlands | poles, general joinery, tannin, cattle food, string and traditional medicines |
| Pseudolachnostylis maprounifolia | Miombo Woodlands | rough carpentry, dye and traditional medicines |
| Pterocarpus rotundifolius | Munga Woodland | tool handles |

*Useful for both fuelwood and charcoal, others for fuelwood only.
Source: Fanshawe (1968)

# 11.     Zimbabwe

The ten biomass classes identified in Zimbabwe can be divided into three broad categories:

i)  Savannah woodlands with a high growing stock and medium levels of productivity. These are found over much of the Highveld and along the mountainous Mozambique border region and are known locally as *Msasa-mnondo* woodlands. Plantations are also included in this category, which accounts for 35.8% of the country. These correspond partially to the closed woody vegetation mapped previously in Zimbabwe. Whitlow (1980a) has measured a 4% decrease in the area under closed woody vegetation (the broad equivalent of this category) between 1963 and 1973.

ii) Savannah woodlands and thickets with medium to high growing stocks and limited annual productivity. These are found on the Highveld, particularly in the north and west, and in the Zambezi Valley. They account for 29.7% of the country. In this zone, which corresponds to the open woody vegetation of other workers, the decrease in areal extent between 1963 and 1973 was 16% (Whitlow 1980a). Lanly (1981) estimated that the combined closed forest and savannah woodland area of Zimbabwe was 190,000 km$^2$ or 51.1% of the country. Equivalent estimates from this study suggest the area covered is slightly higher, accounting for 65.5% of the country.

iii) Low bushy savannah and areas of low bush regrowth in degraded areas with a low growing stock and a low annual level of productivity. These areas are found in the south of the country and in degraded parts of the Highveld. These areas account for 34.5% of Zimbabwe. These types of vegetation generally expanded in area between 1963 and 1973 (Whitlow 1980a).

The distribution of the biomass classes is broadly a function of soils and climate. However, colonial land-zonation policies, which created a division of commercial agricultural land and tribal trust land (now communal land), have also left their mark on the vegetation patterns. The different farming activities in the different zones have altered significantly the natural pattern of growing stocks and productivity. Although land-zonation policy has been transformed since independence, it has not significantly changed the land-use patterns.

With the exception of the more mountainous regions in Manicaland, much of the Highveld was reserved by settlers for commercial farming. Population density on such lands is low and, with the exception of areas around Harare, cultivation is still extensive. Wood resources were destroyed to make way for agricultural land but there is little pressure on the remaining woodlands, some of which are reserved and used for agricultural processing (e.g., tobacco curing). Less fertile areas on the flanks of the Highveld were zoned as Tribal Trust Lands (now Communal Areas or CAs) for African settlement and farming. Cultivation in the north and grazing in the south was, and still is, carried out in the CAs on an intensive basis. Population densities in the CAs are variable, but in most pressure on the renewable natural resource base is high. Many of these CAs now exhibit degraded vegetation with low growing stocks and low annual levels of productivity.

The natural patterns of savannah woodland thickets and bushy savannahs, and the differential population impact on woody-biomass resources due to the land-zonation policies, has led to large regional discrepancies in woody-biomass supply in Zimbabwe. These discrepancies need to be examined on a provincial basis or by assessing the variations between different CAs. Woody-biomass supply can be summarized as follows:

   i)   it is very low in the Matabeleland South;

   ii)   there are serious, but localized, biomass-supply limitations in Masvingo, Matabeleland North, Mashonaland East and Midlands;

   iii)   annual productivity levels are low, but growing stock is generally high, in Mashonaland Central and Mashonaland West; and

   iv)   the only province with few limitations in terms of supply is Manicaland.

This is rather different to the Beijer Institute (1985) estimates, which suggested that four Provinces would experience supply problems by 1987 (Manicaland, Mashonaland East, Masvingo and Midlands). However, these projections compared supply with demand, whereas this report only considers supply data, albeit more up-to-date information than the Beijer Institute assessments. Furthermore, so far in this discussion we have not taken into account the severe access restrictions placed on biomass resources in Manicaland due to forest reservation. Nevertheless it is important to note that the province with the most limited supply, Matabeleland South, was not included in the Beijer Institute projections as a problem area.

The Whitsun Foundation (1981) identified 27 districts as having fuelwood deficits. Initial comparisons between supply data generated in this project and their caluclations show the following trends:

    i)    These districts have low fuelwood supply potential and are identified as deficit areas by the Whitsun Foundation: Bikita-Buhera, Charter, Chilmanzi, Gutu, Gwanda-Insiza, Gweru, Mankoni, Marondera, Mazoe, Mberengwa-Shaba, Mutoko, Mwenezi, Selukwe, Shurugwe and Wedza.

    ii)    The following districts have low fuelwood supply potential but were *not* identified as deficit areas by the Whitsun Foundation: Beitbridge, Bulwayo-Bulalimamangwe, Chiredzi, Harare.

The CAs have also been classified in terms of the biomass-supply data supplied from this study (see Table 11.1).

Zimbabwe is divided into nine biomass classes (see map on p. 179). This division is based on the interpretation of NOAA–7 AVHRR GAC data and reference to previous botanical and forestry studies (Government of Zimbabwe, 1985). The biomass classes are:

    1.    Dense Savannah Woodland
    2.    Open Savannah and *Baikiaea* Woodland
    3.    Seasonal Savannah Woodland
    4.    Dry Savannah Woodland
    5.    Mopane Woodland and Escarpment Thicket
    6.    Dry Bushy Savannah
    7.    Degraded Bushy Savannah
    8.    Degraded Savannah Woodland and Escarpment Thicket
    9.    Intensive Commercial Agriculture

**Table 11.1**  BIOMASS SUPPLY RATING FOR COMMUNAL AREAS IN ZIMBABWE

| | 1 High | | 2 Medium | | 3 Low |
|---|---|---|---|---|---|
| Potential: | High | | Medium | | Low |
| Belingwe | Mtilikwe | Bakasa | Manjolo | Belingwe | Mbongolo |
| (Mberengwe) | Muromo | Bushu | Masembura | (Mberengwe)* | Mondoro |
| Bikita | Musikavanhu | Busi | Masoso | Chibi* | Mtetengwe |
| Chibi* | Mutasa N. | Chete | Matizi | Chikwanda | (Mutoko) |
| Chiduku | Mutasa S. | Chikwakwa | Mkota | Chilimanzi | Mzinyatini |
| Chikukwa | Mutema | Chikwizo | Msana | Chikore | Ngezi |
| Chingowshera | Muwushu | Chimanda | Mudzi | Chiota | Nswazi |
| Chiwindura | Narira | Chinamora | Mukumbura | Dendele | Ntabazinduna |
| Dora | Ndanga | Chinyika | Mukwichi | Debhere | Pfungwe |
| Godhlawyo | Ndowoyo | Chirau | Mzarabani | Dibilishaba | Runde |
| Holdenby | Ngorima | Chiswiti | Mzola | Esirhezini | Sabi* |
| Inkosikazi | Nkai | Chiweshe | Ngarwe | Glassblock | Seearblock |
| Inyanga | Nyajena | Dandana | Nyamaropa | Gulati | Seki |
| Inyati | Rowe | Dande | Omay | Gutu | Selukwe |
| Makoni | Sabi* | Gandavaroyi | Piriwiri | Gwaranyemba | (Shurunguri) |
| Manga | Sabi N. | Gatshe-Gatshe | Rengwe | Insiza | Sengwe |
| Mangwende | Sengwe | Gutsa | St Swithins | Kumalo | Shashi |
| Manjirenyi | Tamandayi | (Hwange) | Sanyati | Manchuchuta | Siyoka |
| Manyika | Wedza | Inyanga N. | Sawunyama | Mambali | Tanda |
| Maranda | Weya | Kachuta | Selukwe | Manyeri | Tshatshani |
| (Masvingo) | Zhombe | Kana | Siabuwa | Maranda | Umzingwane |
| Matsai | Zimunya | Kandeya | Silobela | Maremba | Ungova |
| | | Kanyati | Sipolilo | Marami | Wenlock |
| | | Kunzwi | Tjolotjo | Masera | Zimbali |
| | | Lower Gweru | (Tsholotsho) | (Mashava) | Zimutu |
| | | Lumbimbi | Umfuli | Matibi 1 | |
| | | Lupane | Urungwe | Matibi 2 | |
| | | Madziwa | Uzumba | Matopo | |
| | | Magondi | Zwimba | Matshetshe | |

* Large CAs occurring more or less equally in more than one class.
( ) recent name changes

Biomass Classes

1. High   = 1, 2, 3,
2. Medium = 4, 5,
3. Low    = 6, 7, 8, 9, 10

**Table 11.2** COMPARISON OF BIOMASS SUPPLY IN CAs
WITH ENTIRE COUNTRY

| | *Proportion of CAs in biomass-supply rating class (Table 11.1) (%)* | *Proportion of total land area in equivalent biomass classes (see text) (%)* |
|---|---|---|
| High | 28.76 | 36.64 |
| Medium | 37.25 | 31.06 |
| Low | 33.99 | 32.30 |

The Open Savannah and *Baikiaea* Woodland class also includes montane vegetation.

**DESCRIPTIONS OF INDIVIDUAL BIOMASS CLASSES**

**Dense Savannah Woodland**

Dense Savannah Woodland is scattered amongst the Open Savannah Woodland in the Zimbabwean Highveld. There are four main occurrences of this class: between the Mwenezi and Sabi Rivers in Masvingo Province; between Gweru and the Sabi River in Midlands and Manicaland Provinces; on the watershed between the Sengwa and Shangani Rivers in Mashonaland North and Matabeleland West provinces; and around Harare. The total area covered is 18,907 km² or about 4.8% of Zimbabwe.

This woodland shows only slight variations in productivity throughout the year. Productivity is generally high, with NDVI values exceeding 180 during much of the wet season. It dips slightly in the dry season, reaching its lowest NDVI values of about 170 in September.

This types of woodland is found in moist, undisturbed conditions on well-drained soils, generally above 1,350 masl, but at slightly lower altitudes in the south. The canopy, which is dominated by *Brachystegia spiciformis* and *Julbernardia globiflora*, varies between 6 and 13 m in height. In the slightly wetter eastern areas, *B. utilis* invades the canopy. Tree canopy cover is high, generally over 80%, but the shrub and grass cover is poorly developed and open, usually below 50%. The woodland exhibits less disturbance than the surrounding woodland and there are few, if any, of the grass and savannah areas

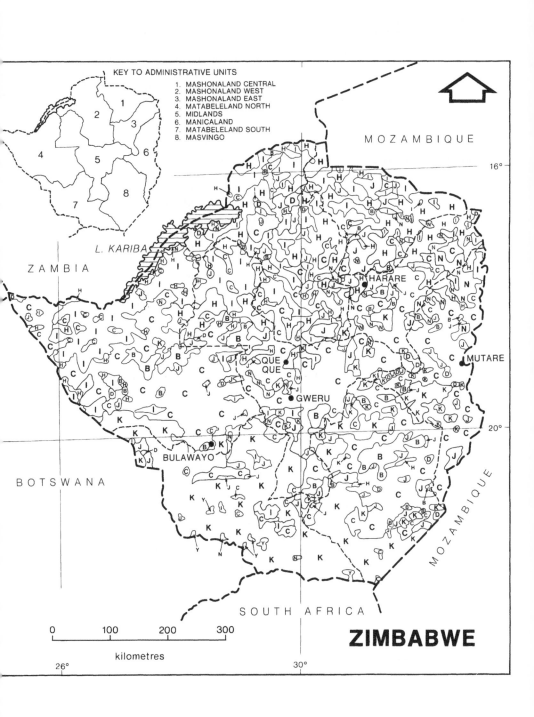

KEY TO ADMINISTRATIVE UNITS

1. MASHONALAND CENTRAL
2. MASHONALAND WEST
3. MASHONALAND EAST
4. MATABELELAND NORTH
5. MIDLANDS
6. MANICALAND
7. MATABELELAND SOUTH
8. MASVINGO

MOZAMBIQUE

16°

L. KARIBA

ZAMBIA

HARARE

MUTARE

QUE
QUE

GWERU

20°

BULAWAYO

BOTSWANA

MOZAMBIQUE

SOUTH AFRICA

| 0 | 100 | 200 | 300 |

kilometres

26°                                              30°

**ZIMBABWE**

that are found in the other types of savannah woodland in Zimbabwe. Overall, the fuelwood potential, in terms of growing stock, is slightly higher than the surrounding woodlands, but the annual productivity is similar.

**Open Savannah and Baikiaea Woodland**

Open Savannah and *Baikiaea* Woodland is the most areally extensive biomass class in Zimbabwe, accounting for 30.2% of the country or 117,790 km². This biomass class dominates the south-eastern Highveld, occurring south of a line running through Harare and Gwasi. It is areally important in Manicaland and Masvingo Provinces and is locally important in Mashonaland East, Mashonaland West, Matebeleland North and Midlands Provinces.

Montane vegetation is also included in this class and is mainly found in the mountainous areas along the Mozambique border, in Manicaland Province. From north to south these are the Inyanga Mountains, the Msere and Shena Hills, the Vumba Mountains, the Chimanimani Mountains, and the Masote and Matzuru Ranges. There are two other relatively large areas near Triangle in Masvingo Province, and near Karoi in Mashonaland West Province. There are other localized occurrences throughout the Highveld savannah woodlands.

Productivity levels are high in the wet season and from January until April, NDVI values range from 170 to 175. Unlike the Dense Savannah Woodland, there is a substantial fall in productivity during the dry season. By August, NDVI values have fallen to 150 and the woodland remains relatively unproductive until November, when productivity begins to rise again. Consequently, it is different to the Dense Savannah Woodland, both in terms of phenology and biomass.

Montane vegetation is dominated by montane forest, woodland and grassland. All of these vegetation types have slightly higher total biomasses than the Open Savannah and *Baikiaea* Woodlands and exhibit a slightly different seasonality. However, neither of these phenomena are different enough to classify them separately at this scale. These forests and woodlands occur in the wetter areas but are nevertheless subject to drought for part of the year. Consequently, they are characterized by high rates of photo-synthetic activity with NDVI values generally greater than 175 and there is a slight, but marked, seasonality. Photosynthetic activity is a little higher between January and March when NDVI values vary between 180 and 190; it then declines slowly to reach the lowest values between August and October (NDVI of about 170–180). Activity then increases again towards the end of the year. The decline in photosynthetic activity, and consequently biomass productivity, is due to the short dry season.

This class is generally found on well-drained soils above 1,350 masl, although it occurs as low as 675 masl on the Limpopo Escarpment. The canopy is between 6–13 m tall and is dominated by *Brachystegia spiciformis* and *Julbernardia globiflora*. The tree canopy varies between 50 – 80%; shrub cover is open and is usually below 50%; and grass cover ranges from 50–80%. A growth in the proportion of *Julbernardia* spp. is associated with increased disturbance, often following agricultural clearance. The shrub and grass cover is generally poor. Within this woodland there are patches of grasslands and savannahs of *Acacia* spp. and *Parinari curatellifolia*. All of these divergent vegetation types may be indicative of disturbance or local environmental conditions. However, the fuelwood potential of these vegetation inclusions is much lower than the *B. spiciformis-J. globiflora* woodland which surround them. In Malawi, the wood of *P. curatellifolia* is noted for charcoal making but this practice has not been recorded in the Zimbabwean literature.

In Matabeleland North, west of Hwange on the deep Kalahari Sands, this biomass class is dominated by an open, dry deciduous *Baikiaea* forest. The open nature of the forest is a result of its exploitation for agriculture, fuelwood and, most importantly, for timber – *Baikiaea plurijuga* is the well-known "Zambezi Teak". The *B. plurijuga* woodland canopy is extremely variable in height, ranging from a dwarf form (reaching heights of only 1–1.5 m) to about 20 m high. Other canopy trees are rare, unless there has been excessive disturbance.

The edges of the *Baikiaea* forests grade into the associated woodlands and bushy savannahs, and there is a mixture of tree species. Associated with *B. plurijuga* on deep, sandy substrates are *Ricinodendron rautanenii, Pterocarpus antunesii, Entandrophragma caudatum, Burkea africana* and *Erythrophelum africanum*. The invasive *Acacia erioloba* and *Combretum collinum* are also widespread. There is a floristically-rich, dense, shrubby under-storey (5–8 m tall) in which the main species are *Baphia massaiensis* subsp. *obovata, Combretum engleri, Dirichetia rogersii, Paropsia brazzeana, Pteleopsis anisoptera* and *Pterocarpus antunesii*. The density of the shrub layer increases with repeated burning. Under such a shrubby under-storey, grasses are poorly developed.

Generally, fuelwood resources in these woodlands are good, with high growing stocks and high productivity levels during the wet season. However, there are problems of low dry-season productivity and areas of low growing stock, in both the Open Savannah and the *Baikiaea* Woodlands.

In Zimbabwe, montane vegetation is complex. True montane forest, similar to that found in Malawi and in East Africa, is rare, for two reasons. First, the Zimbabwean Mountains are generally lower and so the high-altitude, stunted *Ericaceous* forests are found only in the Chimanimani Mountains. Secondly, the wet windward slopes of the Zimbabwean Mountains are in Mozambique while only the dry leeward slopes are in Zimbabwe itself. Consequently, Zimbabwean montane forests suffer more severely from drought stress than the equivalent forest in Mozambique. Because of this, savannah woodland is found at up to 2,100 masl, and only above this altitude are true montane forests and grasslands found. A complex relationship exists between the two. The savannah woodlands extend higher up the slopes where the soils are thin and rocky, and the montane forest extends down the slopes in areas with low soil-moisture deficits. The only area at variance with this zonation is in southern Manicaland, where an area of rain forest is found.

The savannah woodland on the mountain slopes is different from that on the Highveld. The canopy rarely exceeds 6 m in height and is underlain by sparse shrub, fern and grass layers. The canopy is dominated by *Brachystegia spiciformis* with occasional areas of *B. glaucesens, B. taxifolia* and *Uapaca kirkiana*. The main shrub is *Philippia benguelensis;* together with *Brachystegia spiciformis* it dominates the Chimanimani Mountains. Montane forests occur above the savannah woodlands, or in wetter areas at the same altitudes. At these sites, trees rarely exceed 15 m in height and the canopy forms more commonly between 9 and 13 m. A number of smaller trees are found below the canopy which vary in height from 3.5 to 9 m. The most common species are *Agauria salicifolia, Aphloia myrtiflora, Apodytes dimidiata, Cussonia spicata, Diospyros whyteana, Dombeya erythroleuca, Hagenia benguelensis, Ilex mitis, Juniperus procera, Kiggelaria africana, Myrica salicifolia, Nuxia congesta, N. floribunda, Olinia usambarensis, Philippia benguelensis, Pittosporum viridiflorum, Podocarpus latifolius, Prunus africanus, Pterocelastrus* spp., *Rapanea* spp., *Trichocladus ellipticus, Widdringtonia nodiflora* and *Xymalos monospora.* In southern Manicaland, wetter montane forest occurs. This is ecologically similar to the rain forests of Mozambique, and is best developed in the Chirinda

Forest. It is floristically diverse, drawing trees from both the lowland rain forest and the montane forests.

The wet savannah woodland experiences a short period of soil-moisture stress and is therefore slightly deciduous. Its structure is similar to the evergreen woodland with a tree canopy of about 10–13 m in height overlaying poorly developed shrub and grass layers. There is little floristic difference between the savannah woodlands, and it is dominated by *Brachystegia spiciformis* and *Julbernardia globiflora*.

All of the montane woodlands and forests are highly productive and have high growing stocks. With the exception of the rain forest, many of the trees are quite low and can be easily cut for fuelwood. However, these circumstances are made less favourable by the reservation of much of this woodland throughout Zimbabwe, especially in Manicaland.

## Seasonal Savannah Woodland

Seasonal Savannah Woodland is found locally throughout Zimbabwe, but large areas of it are confined to Manicaland and Masvingo Provinces. The total area covered by this biomass class is about 3,144 km², accounting for only 0.8% of the country.

This form of Savannah Woodland is intermediate between the open and dry phases. It occurs mainly at the edge of Open Savannah or montane Woodland and is representative of slightly drier conditions. Its intermediate nature is also seen in its phenology. Photosynthetic activity is high from January to April, when NDVI values are about 185. This is similar to the Open Savannah Woodland and significantly higher than the Dry Savannah Woodland. Productivity begins to decline in May, a month earlier than in the Open Savannah Woodland. Low values of photosynthetic activity occur between July and September. The NDVI values during this period vary from 145 to 150 and are similar to the Dry Savannah Woodland, but are much lower than the wet variant. Activity rises sharply from October onwards. This phenological pattern is indicative of areas with substantial moisture reserves in the wet season and a marked dry-season drought which places the woodland under severe moisture stress for up to three months. Consequently, the woodland is deciduous.

Structurally, the woodland is intermediate between the other types of Savannah Woodland. In less degraded and in moister areas the canopy trees reach 8–10 m and, due to the seasonality, are more open than Open Savannah Woodland. The canopy is dominated by a mixture of typical Zimbabwean savannah woodland species. These are mainly *Brachystegia boehmii, B. spiciformis* and *Julbernardia globiflora*, although other trees are invasive. Tree-canopy cover varies between about 50 and 80%, and the shrub and grass cover is relatively well developed. The under-storey is dominated by shrubs and smaller trees such as *Diospyros kirkii, Faurea saligna, Protea gaguedi, Pseudolachnostylis maprouneifolia* and *Psorospermum febrifugum*.

The woodland exhibits slightly more disturbance than the surrounding Open Savannah and Montane Woodland. The overall fuelwood potential, in terms of growing stock and annual productivity, is slightly lower than that of the surrounding woodlands.

## Dry Savannah Woodland

Dry Savannah Woodland is most extensively developed north of 18°N, in Mashonaland Central and West Provinces. It also occurs on the Charama and Mafungagaubusi Plateaux in Mashonaland West, on the Sabe Escarpment in Masvingo, and as isolated patches in Matabeleland North. The total area of this biomass class in Zimbabwe is about 47,717 km², or 12.2% of the country.

Dry Savannah Woodland has a marked seasonality in photosynthetic activity. Between January and May photosynthetic activity is high with NDVI values ranging up to 175. However, this is lower than in the other types of savannah woodland. Photosynthetic activity falls dramatically in June and July, and NDVI values are about 140 from late July until October. This period of low photosynthetic activity is due to the severe seasonal moisture deficit which is accentuated by the poor water-holding capacities of the soils. Consequently, the deciduous period in the Dry Savannah Woodland can easily last for up to four months.

Dry Savannah Woodland occurs mainly at altitudes of between 1,000 and 1,300 masl, on the Highveld. On the higher ground, it is floristically poor and dominated by *Brachystegia spiciformis* and *Julbernardia globiflora*. Scattered amongst the canopy are smaller trees, mainly *Diospyros kirkii, Faurea saligna, Protea gaguedi, Pseudolachnostylis maprouneifolia* and *Psorospermum febrifugum*. The underlying shrub and grass layers are poorly developed. At slightly lower altitudes, on thinner soils, and on the Zambezi Escarpment, the Dry Savannah Woodland canopy is dominated by *Brachystegia boehmii* although other trees, again mainly *Diospyros kirkii, Faurea salinga, Protea gaguedi, Pseudolachnostylis maprouneifolia* and *Psorospermum febrifugum*, are also present. Below 1,000 m, the *Brachystegia* woodland gives way to a lower canopy woodland (7–13 m) which is dominated by *Julbernardia globiflora*.

Although the canopy of Dry Savannah Woodland often reaches between 10 and 13 m, it can be restricted to as little as 3 m as it becomes drier and more disturbed. If this happens, it takes on the form of an open, shrubby savannah. In such situations the canopy is very disrupted; many shrubs invade and a grass layer develops of 0.6 to 1.2 m in height. In these cases, the main dominant tree is *B. boehmii* but other small trees and shrubs invade such as *B. spiciformis, C. mopane, J. globiflora, Kirkia acuminata* and *Sclerocarya caffra*.

**Mopane Woodland and Escarpment Thicket**

This biomass class is dominated by open *Colophospermum mopane* woodland and the thicket vegetation developed along the Zambezi Escarpment. Mopane Woodland and Escarpment Thicket is found extensively along the borders with Botswana, Zambia and Mozambique. Mopane Woodland is also locally important on the Highveld and in the Limpopo Valley, particularly around Chakari, Chivu, Featherstone, Gweru, Kwekwe and Nkai, and in the Umzingwane Valley. Escarpment Thicket is also found flanking Open Savannah Woodland on the Matopo Hills. Consequently, it is found mainly in Mashonaland Central, Mashonaland East, Matabeleland North, Matabeleland South and Midlands Provinces. It has an areal extent of 68,565 km², accounting for 17.5% of the country's land area.

Despite similarities with the ecological structure of other woodlands and thickets, Mopane Woodland and Escarpment Thicket are phenologically distinct. Wet-season productivity is lower than that of other woodland types, but higher than other varieties of thicket. This means that it has a less marked seasonality. NDVI values in January are similar to those of Dry Miombo Woodland – about 175 to 180 – but they decline slowly from February onwards, falling to about 140 in September and October. This indicates a strong seasonal drought similar to that found in other dry-vegetation communities. The main difference lies in the way in which vegetation die-back commences and the rate at which it proceeds. In Mopane Woodland, die-back starts much earlier and is more steady than in the Dry Savannah Woodland whereas in Escarpment Thicket, it begins later than in the other thicket vegetation types.

The canopy of Mopane Woodland is more open than that of the

other Highveld woodlands. In fact, it is open enough to allow large-scale shrub invasion. In the north, the canopy can attain heights of between 20 and 25 m; more commonly however, it varies between 10 and 15 m. Shrub invasion in drier areas probably means that it can be classed as a scrubby woodland and in these situations, the canopy is unlikely to exceed 5 to 10 m. These areas are classified and discussed in the Dry Bushy Savannah biomass class.

The canopy is dominated by *Colophospermum mopane* but other species frequently found there are *Acacia erioloba*, *A. mellifera (detinens)*, *A. nigrescens*, *Burkea africana*, *Combretum apiculatum*, *C. imberbe*, *C. mossambicense*, *Commiphora mossambicensis*, *Sclerocarya caffra*, *Terminalia prunioides* and *Zizyphus mucronata*. Shrubby species invade the understorey in drier, more open situations; the Mopane Woodland in southern Zimbabwe falls into this category. Shrubs commonly found are *Acacia karroo* (subsp. *heterocantha*). *A. mellifera (detinens)*, *A. tortilis* (subsp. *heterocantha)*, *Commiphora pyracanthoides*, *Dichrostachys cinerea*, *Grewia bicolor*, *G. flava* and *Terminalia sericea*.

Escarpment Thicket occurs along the entire length of the Zambezi Escarpment in northern Zimbabwe. Its vegetation rarely reaches more than 10 m and forms a dense mixture of tree and shrub species. It is dominated by *Combretum* spp., *Commiphora* spp. and *Pterocarpus antunesii*. On drier spurs between river valleys, it grades into the Mopane Woodland.

**Dry Bushy Savannah**  Dry Bushy Savannah is a very important biomass class in southern Zimbabwe. It is found extensively along the hot, dry Limpopo, Mwenezi, Shashe, Tuli and Umzingwane valleys. It also occurs in the headwaters of the Gwai and Nata Valleys, in a south-west/north-east trending stretch of land between the Matopo and Marabada Hills, and as localized areas in the Highveld. It is an areally important biomass class in Masvingo and Matabeland South Provinces; and is locally important in Manicaland, Mashonaland Central, Matabeleland North and Midlands Provinces. In total, it accounts for 91,656 km², or about 23.5% of Zimbabwe's total land area.

The Dry Bushy Savannah biomass class includes a variety of similar vegetation types – the arid and dry mountain bushvelds, the arid spiny and thorn plains bushvelds, *Terminalia* sandveld, and low *Colophospermum mopane* bushland (Werger and Coetzee, 1983). All of the vegetation types are dominated by low trees and bushes, a few emergent trees, and varying amounts of grasses and herbs. The areas are similar in terms of their phenology, biomass productivity and growing stock.

The overall levels of productivity for this class are much lower than those of the savannah woodlands. Also, despite the long dry season, there is relatively little seasonal variation in production. This is because wet-season productivity is very low, with NDVI values varying between 150 and 160. There is a slight decrease in productivity from April onwards but even at the end of the dry season, NDVI values are only about 145.

In the dry valleys of southern Zimbabwe, there are long dry-season soil-moisture deficits. This seasonal drought is accentuated by two factors. First, the soils have low infiltration capacities and run-off rates are high. Consequently, little wet-season rainfall replenishes the soil moisture reserves. Secondly, rooting depth is limited in many cases. However, the vegetation has adjusted to these inhospitable dry-season conditions, and slight variations in soil properties are reflected in equally slight variations in the vegetation communities.

On the alluvial soils of the river valleys, low shrubby vegetation is dominant. Variations in the vegetation communities reflect both changes in soil texture and fertility, as well as disturbance. The latter can be quite extensive, as many of these soils are quite fertile and therefore suitable for grazing or cultivation. A few areas of badly degraded savannah do occur in these valleys and are discussed in the Degraded Bushy Savannah biomass class. The ecological communities range from open grassland with scattered shrubs and emergent trees, to almost continuous low scrubland.

The open, grassy savannah has a low woody-biomass component. This is mainly restricted to scattered emergent trees such as *Acacia* spp., *Dichrostachys cinerea* and *Sclerocarya caffra*, all of which reach heights of about 10 m. There are also scattered shrubs which vary in height from 1 to 3 m and are dominanted by *Grewia flava*, *Ormocarpum trichocarpum*, *Rhus pyroides* and *Zizyphus mucronata*. In other areas, the vegetation is restricted to thorny and spiny shrubs scattered amongst grasslands. The woody shrubs and stunted trees reach about 5 m. Typical woody constituents are *Boscia* spp., *Commiphora* spp., *Dichrostachys cinerea*, *Grewia flava*, *Lycium* spp., *Rhigozum brevispinosum*, *Terminalia prunioides* and *Ziziphus mucronata*. The only common emergent is *Adansonia digitata*. The least woody of these communities are the areas of low, shrubby *Colophospermum mopane*.

Here *C.mopane* rarely exceeds 5 m in height and often reaches only 2 m. It usually occurs as scattered individuals in a shrubby ground-cover with few grasses.

There is a very open savannah of small trees and bushes on the escarpments overlooking the river valleys. The structure varies from dense woody stands reaching 4–8 m in height, through a variety of shrub types, to the most arid form where trees rarely exceed 3 m, and the vegetation is very open with dominant grass and fern layers. The species found here are related to those found in the lowlands, and the main woody species are *Acacia* spp., *Androstachys johnsonii, Boscia albitrunca, Colophospermum mopane, Combretum apiculatum, Commiphora glandulosa, Kirkia acuminata, Pterocarpus rotundifolius, Terminalia prunioides* and *Ximenia americana*. The distribution of the various species is dependent mainly on soil properties.

On the sandy parts of the highveld, *Terminalia* sandveld is found. The canopy in these areas is very open, but the trees are taller than those found in the lowland valleys and on the escarpments. The tree canopy is about 8–10 m high and is underlain by a well-developed grass layer with few shrubs. These areas are grazed extensively and in some places show evidence of overgrazing which is affecting the vegetation structure. The main canopy tree is *Terminalia sericea*, with high proportions of *Burkea africana, Diplorhynchus condylocarpon* and *Pseudolachnostylis maprouneifolia*. Other trees invade but are less important.

Shrubby savannah vegetation is also found on the Highveld, in between areas of savannah woodland. Here, it represents a degraded form of savannah woodland. The vegetation structure and phenology are similar to that in the southern valleys but it is floristically different. Woody species, both trees and shrubs, are composed of elements from the Highveld savannah woodlands and the shrub vegetation to the south.

These areas of shrubby savannah are closely related to some of the CAs. Deforestation is common in the CAs because of fires, land requirements for grazing and cultivation, construction timber, and fuelwood. Particularly important here is the selective cutting of one of the most common Highveld woodland trees, *msasa* (or *B. spiciformis*), for fuelwood. The problem of deforestation is exacerbated by high-population pressure on the woody-biomass resources (Cleghorn, 1966; Furness, 1979; Whitlow, 1980a,b). In the CAs with low woody-biomass resources resulting from deforestation, a positive feedback occurs. The lack of forest cover leads to increased soil erosion, soil fertility declines further still, and the capacity for woodland to regenerate is severely reduced. Two main areas of CAs are affected in this way; they have very low woody-biomass reserves and have therefore been included in this biomass class. These are the area between the Matapos and Marabada Hills, and certain Highveld areas south of Harare (as noted by Whitlow, 1980b). The worst affected CAs are:

i) In the area between the Matapos and Marabada Hills: Chikwanda, Chilimanzi, Denhere, Esirhezini, Glassblock, Gulati, Gutu, Insiza, Kumalo, Mashava, Matopo, Matshetshe, Mzinyatini, Nswani, Runde, Sabe, Selukwe, Sevima, Umzingwane, Zimutu.

ii) In the area to the south of Harare, on the Highveld: Chiota, Manyeri, Mondoro, Ngezi, Seki.

Woody-biomass resources are low in all of the vegetation communities in this biomass class. The existing growing stock is low and the annual productivity level is extremely low, because of the severe seasonal drought and the poor water-holding capacity of many of the soils.

**Degraded Bushy Savannah**

Within the Shashe, Tuli, and Umzingwane Valleys, and to a lesser extent the Limpopo Valley, there are areas where the Dry Bushy Savannah has degraded and it now forms a very low, unproductive scrub and grassland community. Some of these areas are related to settlements (e.g., Antelope and Legion Mines, Kezi, M'phoengs and Tuli); others may relate either to wood exploitation or land clearance, or be due to natural soil and drainage factors. The areas vary in size from about 60 to 400 km² and total 2,380 km² or 0.6% of the country.

The vegetation is floristically similar to the Dry Bushy Savannah, the main difference lying in its lower annual levels of productivity. This is relatively low in the wet season, with NDVI values ranging from about 150 to 155, and peaks at about 160 in April. This is followed by a long, slow decline in productivity until the NDVI values reach their lowest point at about 140 in September and October. Productivity increases sharply after this.

The vegetation mainly consists of very low shrubby *C. mopane*. This is usually about 2 m tall, although it may reach as high as 5 m. It occurs as scattered individuals among a very low shrub ground-cover with few grasses. The fuelwood resources of these areas are severely restricted by the low growing stock and low annual levels of productivity.

**Wooded Grassland**    Areas of Wooded Grassland occur locally throughout the Zimbabwean Highveld. These are related to intensive land clearance for agricultural activity, fuelwood, and timber exploitation. The largest area is found in Manicaland and Mashonaland East Provinces, between the Inyanga Mountains to the south and Shamva to the north. This part of north-east Zimbabwe has been noted as an area at risk from high soil erosion (Stocking & Elwell, 1973), and it has been recently studied in detail by Whitlow (1985). The combined total area is 6,653 km$^2$ or 1.7% of the country.

Productivity levels are very low and strongly seasonal. Photosynthetic activity is highest between February and April when NDVI values range from 170 to 175. Then it declines moderately slowly, reflecting a slow rate of vegetation die-back, until September when the lowest rates of activity occur (NDVI of about 150). This period of low productivity is followed by a relatively sharp increase in photosynthetic activity.

The vegetation is generally low, shrubby thicket and bushland occuring as isolated shrubs, bushes or thickets in grassland and areas of cultivation. On the Highveld it indicates areas of very degraded Dry Savannah Woodland. In other areas it represents equally degraded Mopane Woodland and Escarpment Thicket.

In Manicaland and Mashonaland East, this biomass class is indicative of degraded land in a number of CAs (Chikore, Maramba, parts of Mutoko and Pfungwe, Tanda and Zimbili) but is not exclusive to them. Whitlow (1985) notes that this area is currently characterized by high population pressure leading to bush clearance and cultivation of increasingly marginal land. The area is currently a focus of a number of resettlement schemes. Its distribution is restricted mainly to northern Zimbabwe and it represents the northern equivalent of the degraded savannah woodlands found in the CAs of southern Zimbabwe. The two biomass classes are floristically and structurally similar but have different seasonal patterns of productivity.

Woody-biomass growing stocks and annual productivity levels are very low, and these areas have severely limited fuelwood potential.

**Intensive Commercial Agricultural Land**

Commercial agriculture occurs throughout the Highveld on the land reserved for commercial farming. Areas of intensive commercial farming land occur mainly within the savannah woodland biomass classes but where the level of farming intensity is this great, a separate biomass class was identified. The most extensive areas that have been classified as such extend northwards from Harare to Glendale, and westwards to Bromley. This class covers a total of 33,918 km² or about 8.7% of Zimbabwe.

It has a very distinct phenology which is related more to crop production patterns than natural vegetation growth. Productivity is moderately high throughout the year but there is a slight wet-season peak, particularly between March and April, which is related to the period of maximum crop growth. Productivity declines in the dry season, due to crop senescence and harvesting. Many commercial farms have woodlots of indigenous and exotic trees which have less seasonal productivity, mainly because of high levels of photosynthetic activity during the dry season. The affect of the two types of vegetation on the overall phenology of the class is to create relatively high levels of productivity throughout the year, offsetting the detrimental effect of the dry season on the photosynthetic activity of the crops.

Fuelwood potential in these areas is low for three reasons. First, woodland areas are restricted by widespread maize, cotton, and tobacco cultivation. Secondly, there has been intensive exploitation of the few remaining trees. These are mainly *Brachystegia spiciformis*, *Julbernardia globiflora* and *Uapaca kirkiana* (as well as some introduced exotic tree species). They are used for fuelwood and, to a lesser extent, as constructional timber (Mazambani, 1982). Thirdly, the zonation of much of this land for commercial agriculture during the colonial era has limited access to the remaining woody-biomass resources. This means that at best these areas have a severely limited fuelwood potential for all but the commercial farmers themselves.

**BIOMASS SUPPLY**

Zimbabwe has a total growing stock of 1,505.7 million tonnes and an MAI of 47.55 million tonnes (see Table 11.4). This is about five times that estimated by the Government of Zimbabwe in 1985 (see Table 11.3) which suggests that much higher growing stocks may exist in the country than had previously been considered. The total

**Table 11.4** GEOGRAPHICAL EXTENT OF VEGETATION TYPES, ESTIMATES OF TOTAL AND ACCESSIBLE AREAS SURVIVING IN 1984 AND THEIR GROWING STOCK AND INCREMENT

*(areas in 1,000 ha, growing stock and increment in air dry tonnes)*

| Land Use | Brachystegias | Mopane | Baikiaea | Other Woodland | Grassland | Total | (%) |
|---|---|---|---|---|---|---|---|
| Parks/Wildlife | 716.0 | 2,316.0 | 586.0 | 1,049.0 | 46.0 | 4,695.0 | 12.0 |
| Reserve Forest | 327.0 | 81.0 | 439.0 | 82.0 | – | 929.0 | 2.4 |
| Communal Lands | 6,741.0 | 9,059.0 | 638.0 | 405.0 | 17,791.0 | 45.6 | |
| Commercial Farmland | 6,192.0 | 5,867.0 | 343.4 | 2,502.0 | 721.0 | 15,625.4 | 40.0 |
| Total area | 13,976.0 | 17,323.0 | 1,988.4 | 4,581.0 | 1,172.0 | 39,949.4 | – |
| percent | 35.8 | 44.4 | 5.1 | 11.7 | 3.0 | – | 100.0 |
| | | | | | | | |
| *Estimated Woodland Area Surviving 1984* | | | | | | | |
| Reserve Forest | 320.5 | 79.4 | 395.1 | 80.4 | – | 875.4 | 7.8 |
| Communal Lands | 671.4 | 2,264.8 | 159.5 | 237.0 | – | 3,332.7 | 29.5 |
| Commercial Farmlands | 1,857.6 | 3,520.2 | 206.0 | 1,501.2 | – | 7,085.0 | 62.7 |
| Total area | 2,849.5 | 5,864.4 | 760.6 | 1,818.6 | – | 11,293.1 | 100.0 |
| Less 10 % of original area | 1,397.6 | 1,732.3 | 198.8 | 458.1 | – | 3,786.8 | |
| | | | | | | | |
| *Remaining Accessible Woodland* | 1,451.9 | 4,132.1 | 561.8 | 1,360.5 | | 7,506.3 | |
| Growing Stock: | | | | | | | |
| per ha | 28.0 | 51.6 | 78.4 | 16.6 | – | | |
| total (10⁶ adt) | 40.65 | 213.22 | 44.04 | 22.58 | – | 320.49 | |
| Increment: | | | | | | | |
| per ha per ann | 1.0 | 2.42 | 0.6 | 0.8 | – | | |
| total per ann (10⁶ adt/ann) | 1.45 | 9.99 | 0.34 | 1.09 | – | 12.87 | |

adt × air dried tonnes
Source: Government of Zimbabwe, 1985

**Table 11.4** ZIMBABWE: SUMMARY OF GROWING STOCK AND MAI DATA

| Biomass Class | Area | | Growing Stock | | MAI | |
|---|---|---|---|---|---|---|
| | km² | (%) | (mill t) | (%) | (mill t) | (%) |
| Dense Savannah Woodland | 18,907 | 4.8 | 134.2 | 9.2 | 2.24 | 8.9 |
| Open Savannah and *Baikiaea* Woodland | 117,790 | 30.2 | 839.0 | 55.9 | 26.5 | 55.7 |
| Seasonal Savannah Woodland | 3,144 | 0.8 | 6.2 | 0.4 | 0.2 | 0.3 |
| Dry Savannah Woodland | 47,717 | 12.2 | 45.1 | 3.0 | 1.16 | 2.4 |
| Mopane Woodland and Excarpment Thicket | 68,565 | 17.5 | 253.5 | 16.9 | 7.5 | 15.7 |
| Dry Bushy Savannah | 91,656 | 23.5 | 214.1 | 14.3 | 7.60 | 16.0 |
| Degraded Bushy Savannah | 2,380 | 0.6 | 1.7 | 0.1 | 0.07 | 0.1 |
| Wooded Grassland | 6,653 | 1.7 | 7.6 | 0.5 | 0.36 | 0.8 |
| Intensive Commercial Agriculture | 33,918 | 8.7 | 0 | 0 | 0 | 0 |
| TOTAL | 390,734 | | 1,501.7 | | 47.55 | |

* Montane vegetation is included in the Open Savannah and *Baikiaea* Woodland class

MAI estimated in this study is about 4.5 times as great as the Government of Zimbabwe estimated in 1985 (see Table 11.3). This gives a gross supply rate of 1.65 tonnes per person (based on 1986 population estimates). These figures can be compared to the total estimated fuelwood consumption of 3.12 million tonnes in 1979 (Johnston, 1980), an annual use of only 5.4% of the MAI, and an estimated per capita consumption of 0.07 tonnes per person (Johnston, 1980). Despite these encouraging statistics, the woody-biomass resources are unevenly distributed throughout the country; a fact noted by a number of workers (e.g., Johnson, 1980).

The nine biomass classes that have been identified can be grouped into three broad categories – moist savannah woodlands, dry savannah woodlands and thickets, and bushy savannahs and degraded areas. The first category covers 136,697 km² (approximately 36% of the entire country) and is found in most of the Highveld and along the Mozambique border. The largest class in the group – Open Savannah and *Baikiaea* Woodland – is estimated to possess a growing stock of 838.9 million tonnes and covers an area of 117,790 km², or about 30% of the country. It accounts for over half of the Zimbabwean woody-biomass growing stock (55.9%) and a similar proportion (55.7%) of the MAI (see Tables 11.4 and 11.5). These moderately high levels of productivity are estimated to produce 26.5 million tonnes/year although actual woody-biomass yields are extremely variable, ranging from high amounts through-out the wet season, to substantially lower amounts during the dry season. The growing stocks and productivity of the other moist savannah woodlands are much lower, together accounting for 5.6% of the woodland area. They possess a growing stock almost six times less than that of the Open Savannah and *Baikiaea* Woodlands. Productivity is high, with estimates for MAI of approximately 14.39 million tonnes/year, that is about half as much that of the Open Savannah and *Baikiaea* Woodlands.

If all of the moist woodland classes are grouped together, their dominance of Zimbabwean woody-biomass supplies is obvious. For although they only take up 35.8% of the land area, they possess 65.3% of the woody-biomass growing stock and 64.9% of the MAI.

The growing stock of the drier savannah woodlands (Dry Savannah Woodland, and Mopane Woodland and Escarpment Thicket) which occupy a further 116,282 km² (29.7% of the

**Table 11.5** ZIMBABWE: GROWING STOCK AND MAI DISTRIBUTION BY PROVINCE

| Province | Proportion | |
| --- | --- | --- |
| | Growing Stock (%) | MAI (%) |
| Manicaland | 11.4 | 11.6 |
| Mashonaland C. | 3.1 | 2.9 |
| Mashonaland E. | 6.5 | 6.5 |
| Mashonaland W. | 12.3 | 11.5 |
| Masvingo | 7.9 | 8.3 |
| Matabeleland N. | 25.6 | 25.3 |
| Matabeleland S. | 12.1 | 13.1 |
| Midlands | 21.1 | 20.9 |

country) is estimated to be 298.6 million tonnes with an MAI of 11.6 million tonnes (see Tables 11.4 and 11.5). This group is dominated by the Mopane Woodland and Escarpment Thicket class which is far more productive than the Dry Savannah Woodland. It is particularly interesting to note that the drier savannah woodlands and thickets cover only 6% less of Zimbabwe than the moist savannah woodlands, yet they possess a growing stock that is about 230% lower. Productivity levels are 3.5 times lower than those of the moist savannah woodlands.

Of great importance in southern Zimbabwe is the Dry Bushy Savannah biomass class. It is found extensively along the river valleys and in many of the communal areas. This class accounts for approximately a quarter of the country (23.5%) but the overall productivity levels are considerably lower than those of both the moist and dry savannah woodlands. The MAI is only 7.6 million tonnes (16% of the national total) and the growing stock is 214.1 million tonnes (14.3% of the national total). If all of the bushy savannah and degraded vegetation is grouped together, it accounts for about a third of the country (34.5%) but only 14.9% of the woody-biomass growing stock and 16.9% of the MAI.

However, the natural pattern of vegetation has been significantly modified by colonial land-zonation policies, especially the zonation between TTLs (now CAs) and commercial farming land. The effect this has had on woody-biomass supply has been shown in the general description (see p. 174) and in detail in the Dry Bushy Savannah and Wooded Grassland biomass classes (see p. 178). It is within the CAs, that population density has dramatically exceeded the natural carrying capacity of the land and there has been a rapid depletion of woody-biomass resources. At present, local fuelwood shortages are evident in CAs in Matabeleland South, Masvingo and Midlands Provinces. The only province with very few limitations in terms of woody-biomass supply in the CAs is Manicaland.

Problems of supply also exist because of the selection of specific trees for different end-uses. Some of the most commonly occurring trees in all of the three broad groups of biomass classes are preferred fuelwood species. Trees that are dominant and have been noted by many workers as those preferred for fuelwood and charcoal use in Zimbabwe are:

- *Brachystegia spiciformis*
- *Julbernardia globiflora*
- *Colophospermum mopane*

Other commonly occurring trees are often used for construction purposes. A decline in the number of preferentially selected species, decreasing the quality of the remaining of woodlands (and hence supply), has been noted in CAs in southern Zimbabwe (Du Toit *et al.*, 1984). In the lower-quality degraded woodlands and thickets, the substitution of the less preferred species for fuelwood, such as *Combretum* spp. and *B. boehmii*, has occurred. Similar trends have been noted for other end-uses.

Woody-biomass resources in Zimbabwe are also severely restricted by prohibited access to large tracts of land reserved for forestry operations or game parks. Particular problems occur in Mashonaland West and Matabeleland North Provinces (see Table 11.6).

Zimbabwean woody-biomass resources are poorly distributed with respect to biomass classes. This distortion has been exacerbated by the land zonation policy as well as forest and game park reservation. Particularly critical in terms of woody-biomass supply are:

i) Mashonaland Central, Mashonaland East and Masvingo Provinces, because of low growing stocks and MAIs (see Table 11.4); *and*

ii) Mashonaland West and Matabeleland North, because although there are higher proportions of growing stock and MAI here (see Tables 11.4 and 11.5), large areas of land under restricted access exist (see Table 11.6).

Provinces with generally favourable woody-biomass supply situations are Midlands and Manicaland.

**Table 11.6** WOODY-GROWING STOCK RESTRICTION BY LAND RESERVATION IN ZIMBABWE

| Province | Proportion of woody-biomass growing stock under forest and game reserves |
|---|---|
| Manicaland | 12.3 |
| Mashonaland C. | 0 |
| Mashonaland E. | 8.6 |
| Mashonaland W. | 24.8 |
| Masvingo | 14.0 |
| Matabeleland N. | 46.2 |
| Matabeleland S. | 4.3 |
| Midlands | 12.4 |

# Bibliography

Acocks, J.P.H. (1975). *Veld Types of South Africa*, Botanical Survey of South Africa: Pretoria, Memoir 40 (2nd edn).

Agnew, S. and G.M. Stobbs (1972). *Malawi in Maps*, University of London Press: London.

Alvera, B. (1973). "Estudios en bosques de coniferas del Pirineo Central. Serie A: *Pinar conacebo* de San Juan de la Pena I: Production de hojarsca", *Pirineos*, 109, pp. 17–29.

Banks, P.F. and H.R.R. Metelerkamp (1979). "A Comparison Between the Cost of Growing *Eucalyptus Grandis* for Fuelwood and the Cost of Coal in Zimbabwe Rhodesia", *Suid-Afrikaanse Bosoutydskrif*, 117, pp. 16–18.

Barbosa, L.A. (1970). *Carta fitogeografica de Angola*, Instituto Investigacio Cientifica de Angola: Luanda.

Barreto, L.S. and F.A. Soares (1972). "Carta provisoria de productividade primaria liquida dos esossistermas terrestres de Mocambique", *Rev. Cienc. Agronomicas* (Lourenco Marques) 5-A, pp.11–18.

Beijer Institute (1985). *Policy options for energy and development in Zimbabwe*, Stockholm, vol. 1.

Berry, L. (1971). "Relief and physical features" in L. Berry (ed.), *Tanzania in Maps*, English Universities Press: London, pp. 24–7.

Blair-Rains, A. and A.D. McKay. *The Northern State Lands, Botswana*, LRDC: London, LRDC Land Resource Study, 5.

Celander, N. (1983). *Miombo Woodlands of Africa*, Swedish University of Agricultural Sciences: Uppsala, Working Paper 16.

Chapman, J.D. and F. White (1970). *The Evergreen Forests of Malawi*, Commonwealth Forestry Institute: Oxford.

Child, G.I. and M.J. Duever (1975). *Mineral Cycling in a Tropical Moist Forest Ecosystem*, University of Georgia Press: Athens, USA.

Christensen, B. (1978). "Biomass and primary productivity of *Rhizophora apiculate* VI in a mangrove in southern Thailand", *Aquatic Botany*, 4, pp. 43–52.

Cleghorn, W.B. (1966). "Report on the Conditions of Grazing in Tribal Trust Land", *Rhodesia Agricultural Journal* 63, pp. 57–67.

Coelho, H.V.P. (1967). "Zonagem florestal no Distrito do Cuando Cubango" *Agronomicas Angolana*, Luanda, 26, pp. 3–27.

Coetzee, B.J. and P.J. Nel (1978). "The Sudano–Zambezian region" in M.J.A. Werger (ed), *Biogeography and Ecology of Southern Africa I*, Dr W. Junk: The Hague, pp. 301–462.

Colwell, J.A. (1974). "Vegetation canopy reflectance", *Remote Sensing of Environment*, 3, p. 175.

Compton, R.H. (1966). "Annotated Checklist of the Flora of Swaziland", *Journal of South African Botany*, Suppl, Vol. 6.

Condliffe, I. (1978). *Land Use Plan for Dwambazi Forest Reserve, a Reconnaissance Survey*, Ministry of Agriculture and Natural Resources: Malawi.

Du Toit, R.F., B.M. Campbell, R.A. Haney and D. Dore (1984). *Wood Usage and Tree Planting in Zimbabwe's Communal Lands*, Report for the Forestry Commission of Zimbabwe and the World Bank.

Edmonds, A.C.R. (1976). *Vegetation Maps of Zambia*, 1 : 250, 000.

Edwards, I.D. (1981). "A quantitative description of an area of Indigenous Woodland in the Chikala Hills, Liwonde Forest Reserve", Forestry Research Institute of Malawi: Lilongwe.

Edwards, I.D. (1982). "A quantitative description of an area of Savannah Woodland at Nichira Mountain Conservation Area, near Blantyre", Forestry Research Institute of Malawi: Lilongwe.

Egunjobi, J.K. (1975). "Dry matter production by an immature stand of *Pinus caribea* in Nigeria", *Oikos*, 26, pp. 80–5.

ERL (Energy Resources Ltd), (1985). *A Study of Energy Utilisation and Requirements in the Rural Sector of Botswana*, 2 vols: London.

Fanshawe, D.B. (1962). *Fifty Common Trees of Northern Rhodesia*, Government Printer: Lusaka.

Fanshawe, D.B. (1969). "The vegetation of Zambia", *Forest Research Bulletin*, 7.

FAO (1978). *Perspectives for Forestry Development in Mozambique*: Rome.

FAO (1981). *Evaluacion De Los Recursos Forestales De La Republica Popular De Mozambique*: Rome.

Forestry Research Institute of Malawi (1985). "Fuelwood and Polewood Research Project for the rural population of Malawi", *Forestry Research Record*, 62.

Frolich, A. (1984). "Lesotho" in "Energy and development in Southern Africa. SADCC country studies Part I", *Energy, Environment and Development in Africa 3*, Beijer Institute: Stockholm, pp. 135–170.

Furness, C.K. (1979). "Some aspects of fuelwood usage and consumption in African rural and urban areas in Zimbabwe Rhodesia", *Suid-Afrikaanse Bosboutydskrif*, 117, pp. 10–12.

Gatlin, J.A., R.J. Sullivan and C.J. Tucker (1983). "Monitoring global vegetation using NOAA-AVHRR data", *Proceedings of the IGAARS Symposium San Francisco*, I, PF2, 7.1.

Gay, J. (1984a). *Lesotho village energy survey report*, Aid Resources Report.

Gay J. (1984b).*Lesotho Household Energy Survey – 1984*, Ministry of Cooperatives and Rural Development: Maseru.

Gay, J. and M. Khoboko (1982). *Village Energy Survey Report*, Ministry of Cooperatives and Rural Development: Maseru.

Gillman, C. (1949). "The vegetation-types map of Tanganyika Territory", *Geographical Review*, 39, pp. 7–37.

Goebel, C.J. (1985). *Importance of range and livestock to Lesotho*, Circular RM-4. Research Division, Ministry of Agriculture: Lesotho.

Government of Lesotho (undated). *Third Five Year Development Plan 1980–85*: Maseru.

Government of Malawi (1985). *National Atlas of Malawi*, Lilongwe.

Government of Swaziland (1983). *Timber Statistics for 1982*, Mbabane.

Government of Tanzania (1984). "Fuelwood supply situation for Tanzania for the year 1983", Forest Division: Dar es Salaam.

Guillarmod, A.J. (1968). "Lesotho": in "Conservation of Vegetation of Africa South of the Sahara", *Acta Phytogeographica Suecica*, 54, pp. 253–6.

Guy, G.L. (1970). "*Adansonia digitata* and its rate of growth in relation to rainfall in south central Africa", *Proceedings and Transactions of the Rhodesian Science Association*, 54, pp. 68–84.

Guy, P.R. (1981a). "Changes in the biomass and productivity of woodlands in the Sengwa Wildlife Research Area, Zimbabwe", *Journal of Applied Ecology*, 18, pp. 507–19.

Guy, P.R. (1981b). "The estimation of the above-ground biomass of the trees and shrubs in the Sengwa Wildlife Research Area, Zimbabwe", *Suid-Afrikans. Tydskr. Natuurnar*, 11, pp. 135–42.

Henry, P.W.T. (1978). *Forest inventory and management in the Baikiaea Forest of NE Botswana*, Ministry of Overseas Development (UK), unpublished report.

Herbst S.N. and B.R. Roberts (1974). "The alpine vegetation of the Lesotho Drakensberg: a study in quantitative floristics at Oxbow", *Journal of South African Botany*, 40, pp. 257–67.

Huntley, B.J. (1974). "Vegetation and flora conservation in Angola", *mimeo* (source unknown).

Hursh, C.R. (1960). *The Dry Woodlands of Nyasaland*, International Co-Operation Administration: Salisbury (Harare).

Ingram, C.L. (1984). "The afforestation of dambos and lateritic soils in Malawi, *South African Forestry Journal*, 130, pp. 41–53.

I'ons, J.H. (1967). *Veld types of Swaziland*, Ministry of Agriculture Bulletin, 18, Mbabane.

Jackson, G. (1954). *Preliminary Ecological Survey of Nyasaland*, Proceedings of the Second Inter-African Soils Conference: Leopoldville, Doc. 50, pp. 679–90.

Jackson, G. (1959). *Vegetation Map of Nyasaland* (1 : 1,000,000), Land Husbandry Branch: Lilongwe.

Jackson, G. (1969). "The grasslands of Malawi, Parts I and II", *Society of Malawi Journal*, 22(1), pp. 73–82.

Jelenic, N.E. and J.A. van Vegten, (1981). "A pain in the neck: the firewood situation in South-West Kgatleng, Botswana", *National Institute of Development and Cultural Research*, Research Notes 5, University of Botswana.

Johnston, J.C. (1980). "Wood Fuel: a neglected energy source in Zimbabwe", Energy Symposium 80: Harare.

Justice, C.O. (ed.), (1986). "Monitoring the grasslands of semi-arid Africa using NOAA–AVHRR data", *International Journal of Remote Sensing*, 7(11), pp. 1383–1622.

Keay, J. and A.G. Turton, (1970). "Distribution of biomass and major nutrients in a maritime pine plantation", *Australian Forestry*, 34, pp. 39–48.

Keay, J., A.G. Turton and N.A. Campbell (1970). "Fertilizer response of maritime pine on a lateritic soil", *Australian Forestry*, 33, pp. 248–58.

Kelly, R.D. and B.H. Walker (1976). "The effects of different forms of land-use on the ecology of a semi-arid region in south-eastern Rhodesia", *Journal of Ecology*, 64(2), pp. 553–76.

Kidwell, K.B. (1984). *NOAA Polar Orbital Data Users' Guide (TIROS-N, NOAA-6, 7, 8*, NOAA National Climate Center, Washington DC.

Kriek, W. (1981). "Bossen and Bosbouw in Mozambique", *Nederlands Bosbouw Tijdschrift*, 53(1), pp. 19–27.

Krohn, T. (1977). "Swaziland – contrasts in forestry", *World Forestry*, pp. 534–5.

Laangas, S. (1987). *Weather Satellites for Forest Monitoring?* NORSK Regnesentral Rapport, No. 792: Oslo.

Lanly, J.P. (ed.), (1981). *Tropical Forest Resources Assessment Project (GEMS) Tropical Africa, Tropical Asia, Tropical America*, FAO/UNEP: Rome.

Lanly, J.P. (1983). *Tropical Forest Resources,* FAO Forestry Paper 30, FAO: Rome.

Lees, H.M.N. (1963). *Working Plan for the Forests Supplying the Copperbelt, Western Province,* Government Printer: Lusaka.

McKee, W.H. and E. Shoulders (1974). "Slash pine biomass response to site preparation and soil properties", *Proceedings of the Soil Science Society of America,* 38, pp. 144–8.

Mazambani, D. (1982). "Exploitation of trees around Harare", *Zimbabwe Science News,* 16(1), pp. 253–6.

Menzes, O.J.A. de (1971). "Phyto-ecological study of the Mucope Region and the vegetation map", *Boletim do Instituto Investigacao Cientifica de Angola.* 8(2), pp. 7–53.

Ministry of Agriculture (1979). *Soils of Lesotho. A System of Soil Classification for Interpreting Soil Surveys in Lesotho*: Maseru.

Monteiro, R.F.R. (1970). *Estudo da flora e da vegetacao das florestas abertas do planalto do Bie,* Instituto Investgacio Cientifica de Angola: Luanda.

Monteiro, R.F.R. and R.M.A. Sardinha (1971). *Forest species of Angola. A Study of their Timbers II, Region of Dembos.* Memorias e Trabalhos do Instituto de Investigacao Cientifica de Angola, 1.

Moore, J.E. (1971). "Vegetation" in L. Berry (ed.) *Tanzania in Maps,* English Universities Press: London, pp. 30–31.

Munslow, B. (1984). "Mozambique" in "Energy and development in Southern Africa. SADCC country studies Part II," *Energy, Environment and Development in Africa 4,* Beijer Institute: Stockholm, pp. 5–48.

Murdoch, G. (1968). *Soils and Land Capability in Swaziland,* Ministry of Agriculture: Mbabane.

Norwine, J. and D.H. Greegor (1983). "Vegetation classification based on AVHRR satellite imagery", *Remote Sensing of Environment,* 13, p. 69.

Pedro, J. Gomes and L.A. Grandvaux Barbosa (1955). "A vegetação" in "Esboco do reconhecimento ecólogío–agricola de Moçambique", vol. 2, *Memórias e Trabalhos,* 23, pp. 67–224. Centro de Investigação Cientifica Algodoeira: Lourenço Marques (Maputo).

Persson, R. (1975). *Forest Resources of Africa,* Part I, Country Descriptions, Royal College of Forestry: Stockholm.

Polhill, R. (1968). "Tanzania" in "Conservation of Vegetation in Africa South of the Sahara", *Acta Phytogeographica Suecica,* 54, pp. 166–78.

Pratt, D.J., P.J.P. Greenway, M.D.A. Gwynne (1966). "Classifications of East African Rangeland", *Journal of Applied Ecology* 3/2, pp. 369-82.

Pullinger, J.S. and A.M. Kitchin (1982). *Trees of Malawi*, Blantyre.

Ranger, J. (1978). "Recherches sur les biomasses comparées de deux plantations *Pin laricio* de Corse avec ou sans fertilisation", *Ann. Sci. For.*, 35, pp. 93-115.

Richardson, K. (1983). "Afforestation and Conversation: the Lesotho Woodlot Project" in G. Schmitz (ed.), *Lesotho Environment and Management*, Morija: Morija Printing Works.

Richardson, K.F. (1984). "Afforestation of adverse sites: recent improvements in the planting technique in Lesotho", *South African Forestry Journal*, 130, pp. 19-25.

Ringrose, S. and W. Matheson (1988). "Monitoring desertification in Botswana using Landsat MSS data: with consideration as to the nature of the infra-red paradox" in A.C. Millington, S.K. Mutiso and J.A. Binns (eds), *African Resources*, Volume I: "Appraisal and Monitoring", Reading Geographical Papers, 96, pp. 13-26.

Russell, E.W. (ed.), (1962). *The Natural Resources of East Africa*, East African Literature Bureau: Nairobi.

Rutherford, M.C. (1982). "Woody plant biomass distribution in *Burkea africana* savannas" in *Ecology of Tropical Savannas* (B.J. Huntley and P.H. Walker, eds) Springer-Verlag: Heidelberg.

Shaxson, T.F. (1977). "A map of the distribution of major biotic communities", *Society of Malawi Journal*, 30, pp. 36-48.

Schneider, S.R., S.R. McGinnis Jr and J.A. Gatlin (1981). "Use of NOAA/AVHRR visible and near-infra red data for land remote sensing", *NOAA Technical Report, NESS84*, USDC: Washington, DC.

Scobey, R. (1984). "Malawi" in "Energy and development in Southern Africa. SADCC country studies Part I", *Energy, Environment and Development in Africa 3*, Beijer Institute: Stockholm, pp. 171-186.

Soares, F.A. and L.S. Barreto (1972). "Ecological zonation of Mozambique according to the Holdridge System, *Revista de Ciencias Agronomicas* Mozambique, 5, pp. 29-39.

Steele, R. and T. Ncholu (1983). *Woodlot Marketing Study (North) Final Report*, Ministry of Agriculture: Maseru.

Stocking, M.A. and H.A. Elwell (1973). "Soil erosion hazard in Rhodesia", *Rhodesia Agricultural Journal*, 70(4), pp. 93-101.

Stomgaard, P. (1985a). "Biomass estimation equations for miombo woodland, Zambia", *Agroforestry Systems*, 3(1), pp. 3–13.

Stomgaard, P. (1985b). "Biomass, growth and burning of woodland in a shifting cultivation area of south central Africa", *Forest Ecology and Management*, 12, pp. 163–78.

Stomgaard, P. (1986). "Early secondary succession in abandoned shifting cultivator's plots in the Miombo of south central Africa", *Biotropica*, 18(2), pp. 97–106.

Tarpley, J.D., S.R. Schneider and R.L. Money (1984). "Global vegetation indices from NOAA-7 meteorological satellite", *Journal of Climatology and Applied Meteorology*, 23, p. 491.

Townshend, J.R.G., C.O. Justice and V. Kalb (1987). "Characterization and classification of South American Land Cover types using satellite data", *International Journal of Remote Sensing*, 8(8), pp. 1189–1207.

Townshend, J.R.G. and C.J. Tucker (1984). "Objective assessment of AVHRR data for land cover mapping", *International Journal of Sensing*, 6, p. 127.

Trapnell, C.G. (1953). *The Soils, Vegetation and Agriculture of North eastern Rhodesia*, Government Printer: Lusaka.

Trapnell, C.G. and J.N. Clothier (1957). *The Soils, Vegetation and Agriculture of North western Rhodesia*, Government Printer, Lusaka.

Trapnell, C.G., J.D. Martin and W. Allen (1952). *Vegetation-Soil Map of Northern Rhodesia* (1 : 1,000,000), Lusaka.

Tucker, C.J., C. Vanpraet, E. Boerwinkel and A. Gaston (1983). "Satellite sensing of total dry matter accumulation in the Senegalese Sahel", *Remote Sensing of Environment*, 13, p. 461.

Tucker, C.J., J.A. Gatlin, S.R. Scheider and A. Kuchinos (1984a). "Monitoring vegetation in the Nile Valley with NOAA-6 and NOAA-7 AVHRR", *Photogrammetric Engineering and Remote Sensing*, 50, p. 53.

Tucker, C.J., B.N. Holben and T.E. Goff (1984b). "Intensive forest clearing in Rondonia, Brazil as detected by satellite remote sensing" *Remote Sensing of Environment*, 15, p. 255.

Tucker, C.J., J.R.G. Townshend and T.E. Goff (1985a). "African land cover classification using satellite data", *Science*, 227, p. 110.

Tucker, C.J., C.L. Vanpraet, M.J. Sharman and G. Van Ittersum (1985b). "Satellite remote sensing of total herbaceous biomass production in the Senegalese Sahel: 1980–84", *Remote Sensing of Environment*, 17, p. 233.

Tucker, C.J., I.Y. Fung, C.D. Keeling and R.H. Gammon (1986). "Relationship between atmospheric $CO_2$ variations and a satellite-derived vegetation index", *Nature* (London), 319, p. 195.

Turton, A.G. and J. Keay (1970). "Changes in dry weight and nutrient distribution in maritime pine after fertilization", *Australian Forestry*, 34, pp. 84–96.

Vesey-Fitzgerald, D.F. (1963). "Central African grassland", *Journal of Ecology*, 51, pp. 243–74.

Weare, P.R. and A. Yalala (1971). "Provisional vegetation map of Botswana", *Botswana Notes and Records*, 3.

Werger, M. and D. Coetzee (1983). *The Biogeography of Southern Africa*, 2 vols, Dr W. Junk, Amsterdam.

Werger, M.J.A. and B.J. Coetzee (1978). "The Sudano–Zambezian region" in M.J.A. Werger (ed.), *Biogeography and Ecology of Southern Africa I*, Dr W. Junk: The Hague, pp. 301–462.

White, F. (1965). "The savanna woodlands of the Zambesian and Sudanian domains", *Webbia*, 19, pp. 651–81.

White, F. (1983). *The Vegetation of Africa*, UNESCO Natural Resources Research XX: Paris.

White, R. (1979). *A Woodlot Management Plan for Martsheng Villages*, Government of Botswana (Village Area Development Project), Hukuntsi: Gaborone.

Whitlow, J.R. (1980a). *Deforestation in Zimbabwe – Problems and Prospects*, Zambezia Supplement.

Whitlow, J.R. (1980b). *Deforestation in Zimbabwe – Some Problems and Prospects*, Government Printer: Harare.

Whitlow, J.R. (1985). An erosion survey of the Mutoko Region in north-east Zimbabwe, *Zimbabwe Agricultural Journal*, 82.

Whittaker, R.H. and W.A. Niering (1975). "Vegetation of the Santa Catalina Mountains, Arizona: V: Biomass, production and diversity along the elevation gradient", *Ecology*, 56, pp. 771–90.

Whitsun Foundation (1981). *Rural Afforestation Study*, Harare.

Wickstead, M. (1984). *Marketing Study (South)*, Ministry of Agriculture: Maseru.

World Bank (1980). *Staff Appraisal Report: Malawi Wood Energy Project*, Report No. 2625–MAI.

Young, A. and P. Brown (1962). *The Physical Environment of Northern Nyasaland, with special reference to soils and agriculture*, Government Printer, Zomba.

# Woody Biomass Supply Statistics

## ANGOLA
### Bengo and Luanda Provinces

| Land Cover Type | Area (km²) | Growing Stock (mill t) | Mean Annual Increment (mill t) |
|---|---|---|---|
| Dense, Medium-Height Miombo Woodland | 11,274 | 80.29 | 2.54 |
| Seasonal Miombo Woodland and Wooded Savannah | 1,478 | 2.93 | 0.07 |
| Dry Coastal Savannah and Arid Coastal Thicket | 19,901 | 23.54 | 0.84 |
| Degraded Miombo Woodland | 1,147 | 1.08 | 0.03 |
| TOTAL | 33,800 | 107.84 | 3.48 |

### Bie Province

| Land Cover Type | Area (km²) | Growing Stock (mill t) | Mean Annual Increment (mill t) |
|---|---|---|---|
| Transitional Rain Forest/Miombo Woodland | 12,851 | 91.52 | 2.89 |
| Dense, High Miombo Woodland | 10,520 | 74.92 | 2.37 |
| Dense, Medium-Height Miombo Woodland | 23,410 | 166.73 | 5.27 |
| Seasonal Miombo Woodland and Wooded Savannah | 24,091 | 47.82 | 1.18 |
| Dry Deciduous Savannah | 1,208 | 1.73 | 0.05 |
| TOTAL | 71,900 | 382.72 | 11.76 |

**Benguela Province**

| Land Cover Type | Area (km²) | Growing Stock (mill t) | Mean Annual Increment (mill t) |
|---|---|---|---|
| Transitional Rain Forest/ Miombo Woodland | 180 | 1.28 | 0.04 |
| Dense, High Miombo Woodland | 180 | 1.28 | 0.04 |
| Dense, Medium-Height Miombo Woodland | 5,670 | 40.38 | 1.28 |
| Seasonal Miombo Woodland and Wooded Savannah | 12,603 | 25.02 | 0.62 |
| Dry Deciduous Savannah | 180 | 0.30 | 0.01 |
| Dry Coastal Savannah and Arid Coastal Thicket | 14,794 | 17.50 | 0.62 |
| Degraded Miombo Woodland | 1,727 | 1.63 | 0.04 |
| Coastal and Desert Vegetation | 3,866 | 0 | 0 |
| TOTAL | 39,200 | 87.39 | 2.65 |

**Cunene Province**

| Land Cover Type | Area (km²) | Growing Stock (mill t) | Mean Annual Increment (mill t) |
|---|---|---|---|
| Dense High Miombo Woodland | 5,516 | 34.71 | 1.10 |
| Dense, Medium-Height, Miombo Woodland | 12,230 | 87.10 | 2.75 |
| Dry Deciduous Savannah | 49,162 | 82.54 | 2.41 |
| Dry Inland Savannah | 2,758 | 3.26 | 0.12 |
| Degraded Miombo Woodland | 2,878 | 2.72 | 0.07 |
| Degraded Dry Deciduous Savannah | 4,556 | 10.64 | 0.38 |
| TOTAL | 77,100 | 141.83 | 4.03 |

## Cabinda Province

| Land Cover Type | Area (km²) | Growing Stock (mill t) | Mean Annual Increment (mill t) |
|---|---|---|---|
| Transitional Rain Forest/Miombo Woodland | 400 | 2.85 | 0.09 |
| Dense, High Miombo Woodland | 400 | 2.85 | 0.09 |
| Seasonal Miombo Woodland and Wooded Savannah | 3,200 | 6.35 | 0.16 |
| Dry Coastal Savannah | 400 | 0.47 | 0.02 |
| Degraded Rain Forest | 1,800 | 1.70 | 0.04 |
| Coastal Vegetation | 900 | 0 | 0 |
| TOTAL | 7,100 | 14.22 | 0.40 |

## Huambo Province

| Land Cover Type | Area (km²) | Growing Stock (mill t) | Mean Annual Increment (mill t) |
|---|---|---|---|
| Transitional Rain Forest/ Miombo Woodland | 4,873 | 34.71 | 1.10 |
| Dense, High Miombo Woodland | 974 | 6.94 | 0.22 |
| Dense Medium-Height Miombo Woodland | 9,422 | 67.10 | 2.12 |
| Seasonal Miombo Woodland and Wooded Savannah | 12,281 | 24.38 | 0.60 |
| Dry Deciduous Savannah | 4,873 | 8.18 | 0.24 |
| Degraded Miombo Woodland | 552 | 0.52 | 0.01 |
| Montane Grassland | 325 | 0 | 0 |
| TOTAL | 33,300 | 141.83 | 4.03 |

## Huila Province

| Land Cover Type | Area (km²) | Growing Stock (mill t) | Mean Annual Increment (mill t) |
|---|---|---|---|
| Transitional Rain Forest/Miombo Woodland | 508 | 3.62 | 0.11 |
| Dense, High Miombo Woodland | 3,563 | 25.38 | 0.80 |
| Dense, Medium-Height, Miombo Woodland | 44,789 | 318.99 | 10.08 |
| Seasonal Miombo Woodland and Wooded Savannah | 9,619 | 19.09 | 0.47 |
| Dry Deciduous Savannah | 16,287 | 27.35 | 0.80 |
| Degraded Miombo Woodland | 4,326 | 4.08 | 0.10 |
| Montane Grassland | 508 | 0 | 0 |
| TOTAL | 79,600 | 398.51 | 12.36 |

## Kuando Kubango Province

| Land Cover Type | Area (km²) | Growing Stock (mill t) | Mean Annual Increment (mill t) |
|---|---|---|---|
| Dense, High Miombo Woodland | 72,429 | 515.84 | 16.30 |
| Dense, Medium-Height Miombo Woodland | 11,043 | 78.65 | 2.48 |
| Dry Deciduous Savannah | 87,044 | 146.15 | 4.27 |
| Degraded Dry Deciduous Savannah | 2,276 | 5.32 | 0.19 |
| Chanas da Borracha Grassland | 26,308 | 0 | 0 |
| TOTAL | 199,100 | 745.96 | 23.24 |

## Kwanza Norte Province

| Land Cover Type | Area (km²) | Growing Stock (mill t) | Mean Annual Increment (mill t) |
|---|---|---|---|
| Dense, Medium-Height Miombo Woodland | 7,139 | 50.84 | 1.61 |
| Seasonal Miombo Woodland and Wooded Savannah | 14,635 | 29.05 | 0.72 |
| Dry Inland Savannah | 5,354 | 6.33 | 0.22 |
| Degraded Rain Forest | 1,071 | 1.01 | 0.03 |
| TOTAL | 28,199 | 87.23 | 2.58 |

## Kwanza Sul Province

| Land Cover Type | Area (km²) | Growing Stock (mill t) | Mean Annual Increment (mill t) |
|---|---|---|---|
| Transitional Rain Forest/Miombo Woodland | 4,741 | 33.77 | 1.07 |
| Dense, Medium-Height Miombo Woodland | 5,911 | 42.10 | 1.33 |
| Seasonal Miombo Woodland and Wooded Savannah | 34,993 | 69.46 | 1.71 |
| Dry Coastal Savannah and Arid Coastal Thicket | 8,946 | 10.58 | 0.38 |
| Degraded Miombo Woodland | 2,845 | 2.69 | 0.07 |
| Coastal Vegetation | 1,264 | 0 | 0 |
| TOTAL | 58,700 | 158.60 | 4.56 |

## Lunda Norte Province

| Land Cover Type | Area (km²) | Growing Stock (mill t) | Mean Annual Increment (mill t) |
|---|---|---|---|
| Transitional Rain Forest/Miombo Woodland | 67,499 | 480.73 | 15.19 |
| Dense, Medium-Height Miombo Woodland | 2,984 | 21.25 | 0.67 |
| Seasonal Miombo Woodland and Wooded Savannah | 31,698 | 62.92 | 1.55 |
| Degraded Rain Forest | 1,119 | 1.06 | 0.03 |
| TOTAL | 103,300 | 565.96 | 17.44 |

## Lunda Sul Province

| Land Cover Type | Area (km²) | Growing Stock (mill t) | Mean Annual Increment (mill t) |
|---|---|---|---|
| Transitional Rain Forest/Miombo Woodland | 5,740 | 40.88 | 1.29 |
| Dense, Medium-Height Miombo Woodland | 19,012 | 135.40 | 4.28 |
| Seasonal Miombo Woodland and Wooded Savannah | 43,047 | 85.45 | 2.11 |
| *Chanas da Borracha* Grassland | 3,101 | 0 | 0 |
| TOTAL | 70,900 | 261.73 | 7.68 |

## Malanje Province

| Land Cover Type | Area (km²) | Growing Stock (mill t) | Mean Annual Increment (mill t) |
|---|---|---|---|
| Transitional Rain Forest/Miombo Woodland | 26,744 | 190.47 | 6.02 |
| Dense, High Miombo Woodland | 508 | 3.62 | 0.11 |
| Dense, Medium-Height Miombo Woodland | 8,328 | 59.31 | 1.87 |
| Seasonal Miombo Woodland and Wooded Savannah | 50,781 | 100.80 | 2.49 |
| Degraded Rain Forest | 2,539 | 2.40 | 0.06 |
| TOTAL | 88,900 | 356.60 | 10.55 |

## Moxico Province

| Land Cover Type | Area (km²) | Growing Stock (mill t) | Mean Annual Increment (mill t) |
|---|---|---|---|
| Transitional Rain Forest/Miombo Woodland | 2,641 | 18.81 | 0.59 |
| Dense High Miombo Woodland | 15,837 | 112.79 | 3.56 |
| Dense, Medium-Height Miombo Woodland | 48,098 | 342.55 | 10.82 |
| Seasonal Miombo Woodland and Wooded Savannah | 30,793 | 61.12 | 1.51 |
| Dry Deciduous Savannah | 68,334 | 114.73 | 3.35 |
| Degraded Dry Deciduous Savannah | 28,155 | 65.77 | 2.34 |
| *Chanas da Borracha* Grassland | 7,842 | 0 | 0 |
| TOTAL | 201,700 | 715.77 | 22.17 |

**Namibe Province**

| Land Cover Type | Area (km²) | Growing Stock (mill t) | Mean Annual Increment (mill t) |
|---|---|---|---|
| Dense, High Miombo Woodland | 687 | 4.89 | 0.15 |
| Dense, Medium-Height Miombo Woodland | 1,457 | 10.38 | 0.33 |
| Seasonal Miombo Woodland and Wooded Savannah | 1,843 | 3.66 | 0.09 |
| Dry Deciduous Savannah | 2,750 | 4.62 | 0.13 |
| Dry Inland Savannah | 18,151 | 21.47 | 0.76 |
| Degraded Miombo Woodland | 1,237 | 1.17 | 0.03 |
| Bushy Arid Shrubland | 15,748 | 11.20 | 0.47 |
| Desert Vegetation | 15,127 | 0 | 0 |
| TOTAL | 57,000 | 57.39 | 1.96 |

**Uige Province**

| Land Cover Type | Area (km²) | Growing Stock (mill t) | Mean Annual Increment (mill t) |
|---|---|---|---|
| Transitional Rain Forest/Miombo Woodland | 33,340 | 237.52 | 7.50 |
| Dense, High Miombo Woodland | 667 | 4.75 | 0.15 |
| Dense, Medium-Height Miombo Woodland | 2,000 | 14.24 | 0.45 |
| Seasonal Miombo Woodland and Wooded Savannah | 17,003 | 33.75 | 0.83 |
| Degraded Miombo Woodland | 3,290 | 3.11 | 0.08 |
| Degraded Rain Forest | 3,000 | 2.83 | 0.07 |
| TOTAL | 59,300 | 296.20 | 9.08 |

## Zaire Province

| Land Cover Type | Area (km²) | Growing Stock (mill t) | Mean Annual Increment (mill t) |
|---|---|---|---|
| Transitional Rain Forest/Miombo Woodland | 163 | 1.16 | 0.04 |
| Dense, Medium-Height Miombo Woodland | 8,397 | 59.80 | 1.89 |
| Seasonal Miombo Woodland and Wooded Savannah | 18,881 | 37.48 | 0.93 |
| Dry Coastal Savannah | 4,443 | 5.26 | 0.19 |
| Degraded Rain Forest | 5,689 | 5.37 | 0.14 |
| Coastal Vegetation | 27 | 0 | 0 |
| TOTAL | 37,600 | 109.07 | 3.99 |

## BOTSWANA
### Barolong and South East Districts

| Land Cover Type | Area (km²) | Growing Stock (mill t) | Mean Annual Increment (mill t) |
|---|---|---|---|
| Bushland with Scrubby Woodland and Woody Shrubland | 315 | 0.45 | 0.02 |
| Shrubland and Bushy Shrubland | 2,565 | 1.82 | 0.08 |
| TOTAL | 2,880 | 2.27 | 0.10 |

## Central District

| Land Cover Type | Area (km²) | Growing Stock (mill t) | Mean Annual Increment (mill t) |
|---|---|---|---|
| Dense Woodland | 3,916 | 27.89 | 0.88 |
| Open Woodland | 9,557 | 35.33 | 1.04 |
| Woodland and Bushland | 78,783 | 184.04 | 6.54 |
| Bushland with Scrubby Woodland and Woody Shrubland | 27,818 | 39.56 | 1.67 |
| Hill Shrubland and Woodland | 4,684 | 5.36 | 0.26 |
| Shrubland and Bushy Shrubland | 3,635 | 2.59 | 0.11 |
| Salt Pans | 14,228 | 0 | 0 |
| Fringing Palm Woodland | 3,115 | 2.21 | 0.09 |
| FOREST/GAME RESERVES | | | |
| Open Woodland | 866 | 3.21 | 0.09 |
| Shrubland & Bushy Shrubland | 1,189 | 0.85 | 0.04 |
| TOTAL | 147,791 | 301.03 | 10.72 |

## Kgatleng District

| Land Cover Type | Area (km²) | Growing Stock (mill t) | Mean Annual Increment (mill t) |
|---|---|---|---|
| Bushland with Scrubby Woodland and Woody Shrubland | 2,977 | 4.23 | 0.18 |
| Shrubland and Bushy Shrubland | 4,983 | 3.54 | 0.15 |
| TOTAL | 7,960 | 7.77 | 0.33 |

## Chobe District

| Land Cover Type | Area (km²) | Growing Stock (mill t) | Mean Annual Increment (mill t) |
|---|---|---|---|
| Dense Woodland | 2,228 | 15.87 | 0.50 |
| Open Woodland | 6,554 | 24.23 | 0.71 |
| Hill Shrubland and Woodland | 344 | 0.39 | 0.02 |
| Salt Pans | 2,168 | 0 | 0 |
| FOREST/GAME RESERVES | | | |
| Dense Woodland | 8,913 | 63.48 | 2.01 |
| Open Woodland | 592 | 2.19 | 0.06 |
| TOTAL | 20,899 | 106.16 | 3.3 |

## Kweneng District

| Land Cover Type | Area (km²) | Growing Stock (mill t) | Mean Annual Increment (mill t) |
|---|---|---|---|
| Woodland and Bushland | 6,760 | 15.79 | 0.56 |
| Bushland with Scrubby Woodland and Woody Shrubland | 6,870 | 9.77 | 0.41 |
| Shrubland and Bushy Shrubland | 16,282 | 11.58 | 0.49 |
| Salt Pans | 550 | 0 | 0 |
| FOREST/GAME RESERVES | | | |
| Shrubland and Bushy Shrubland | 5,428 | 3.86 | 0.16 |
| TOTAL | 35,890 | 41.00 | 1.62 |

## Ghanzi District

| Land Cover Type | Area (km²) | Growing Stock (mill t) | Mean Annual Increment (mill t) |
|---|---|---|---|
| Dense Woodland | 138 | 0.98 | 0.03 |
| Open Woodland | 7,378 | 27.28 | 0.80 |
| Woodland and Bushland | 37,713 | 88.10 | 3.13 |
| Bushland with Scrubby Woodland and Woody Shrubland | 20,088 | 28.57 | 1.21 |
| Hill Shrubland and Woodland | 10,990 | 12.58 | 0.60 |
| Shrubland and Bushy Shrubland | 1,250 | 0.89 | 0.04 |
| FOREST/GAME RESERVES | | | |
| Open Woodland | 4,328 | 16.00 | 0.47 |
| Woodland and Bushland | 25,142 | 58.73 | 2.09 |
| Bushland with Scrubby Woodland and Woody Shrubland | 8,555 | 12.17 | 0.51 |
| Shrubland and Bushy Shrubland | 2,322 | 1.65 | 0.07 |
| TOTAL | 117,904 | 246.95 | 8.95 |

## North-East District

| Land Cover Type | Area (km²) | Growing Stock (mill t) | Mean Annual Increment (mill t) |
|---|---|---|---|
| Dense Woodland | 114 | 0.81 | 0.03 |
| Open Woodland | 114 | 0.42 | 0.01 |
| Woodland and Bushland | 4,892 | 11.43 | 0.41 |
| TOTAL | 5,120 | 12.66 | 0.45 |

## Kgalagadi Province

| Land Cover Type | Area (km²) | Growing Stock (mill t) | Mean Annual Increment (mill t) |
|---|---|---|---|
| Woodland and Bushland | 21,139 | 49.38 | 1.75 |
| Bushland with Scrubby Woodland and Woody Shrubland | 10,605 | 15.08 | 0.64 |
| Hill Shrubland and Woodland | 705 | 0.81 | 0.04 |
| Shrubland and Bushy Shrubland | 31,426 | 22.34 | 0.94 |
| Salt Pans | 2,419 | 0 | 0 |
| FOREST/GAME RESERVES | | | |
| Bushland with Scrubby Woodland and Woody Shrubland | 19,695 | 28.01 | 1.18 |
| Shrubland and Bushy Shrubland | 20,952 | 14.90 | 0.63 |
| TOTAL | 106,941 | 130.52 | 5.18 |

## Ngwaketse District

| Land Cover Type | Area (km²) | Growing Stock (mill t) | Mean Annual Increment (mill t) |
|---|---|---|---|
| Open Woodland | 31 | 0.11 | <0.01 |
| Woodland and Bushland | 4,344 | 10.15 | 0.36 |
| Bushland with Scrubby Woodland and Woody Shrubland | 7,954 | 11.31 | 0.48 |
| Shrubland and Bushy Shrubland | 13,828 | 9.83 | 0.41 |
| Salt Pans | 612 | 0 | 0 |
| Hill Shrubland and Woodland | 612 | 0.70 | 0.03 |
| TOTAL | 27,381 | 32.10 | 1.29 |

**Ngamiland District**

| Land Cover Type | Area (km²) | Growing Stock (mill t) | Mean Annual Increment (mill t) |
|---|---|---|---|
| Dense Woodland | 18,052 | 128.57 | 4.06 |
| Open Woodland | 48,816 | 180.47 | 5.32 |
| Woodland and Bushland | 18,891 | 44.13 | 1.57 |
| Riparian Woodland | 2,204 | 15.70 | 0.50 |
| Bushland with Scrubby Woodland and Woody Shrubland | 1,866 | 2.65 | 0.11 |
| Hill Shrubland and Woodland | 9,167 | 10.50 | 0.50 |
| Shrubland and Bushy Shrubland | 233 | 0.17 | 0.07 |
| Salt Pans | 607 | 0 | 0 |
| FOREST/GAME RESERVES | | | |
| Open Woodland | 5,201 | 19.23 | 0.57 |
| Riparian Woodland | 4,093 | 29.15 | 0.92 |
| TOTAL | 109,130 | 430.56 | 13.62 |

## LESOTHO
### Berea District

| Land Cover Type | Area (km²) | Growing Stock (1,000 tonnes) | Mean Annual Increment (1,000 tonnes) |
|---|---|---|---|
| Escarpment and Riparian Woodland | 115 | 161.46 | 1.72 |
| Escarpment Grassland with Scrub Woodland | 574 | 28.70 | 0.57 |
| Highveld and Riparian Grassland | 408 | 0 | 0 |
| Alpine/Sub-Alpine Grassland and Heathland | 765 | 0 | 0 |
| Small farm cultivation | 360 | 18.00 | 0.36 |
| TOTAL | 2,222 | 208.16 | 2.65 |

### Butha Buthe District

| Land Cover Type | Area (km²) | Growing Stock (1,000 tonnes) | Mean Annual Increment (1,000 tonnes) |
|---|---|---|---|
| Escarpment and Riparian Woodland | 284 | 398.74 | 4.26 |
| Escarpment Grassland with Scrub Woodland | 407 | 20.35 | 0.41 |
| Alpine/Sub-Alpine Grassland and Heathland | 917 | 0 | 0 |
| Small farm cultivation | 105 | 5.25 | 0.10 |
| TOTAL | 1,767 | 424.34 | 4.77 |

**Leribe District**

| Land Cover Type | Area (km²) | Growing Stock (1,000 tonnes) | Mean Annual Increment (1,000 tonnes) |
|---|---|---|---|
| Escarpment and Riparian Woodland | 345 | 484.38 | 5.17 |
| Escarpment Grassland with Scrub Woodland | 448 | 22.40 | 0.45 |
| Highveld and Riparian Grassland | 149 | 0 | 0 |
| Alpine/Sub-Alpine Grassland and Heathland | 1,398 | 0 | 0 |
| Small farm cultivation | 488 | 24.40 | 0.49 |
| TOTAL | 2,828 | 531.18 | 6.11 |

**Mafeteng District**

| Land Cover Type | Area (km²) | Growing Stock (1,000 tonnes) | Mean Annual Increment (1,000 tonnes) |
|---|---|---|---|
| Escarpment Grassland with Scrub Woodland | 85 | 4.25 | 0.09 |
| Highveld and Riparian Grassland | 1,221 | 0 | 0 |
| Alpine/Sub-Alpine Grassland and Heathland | 90 | 0 | 0 |
| Small farm cultivation | 623 | 31.15 | 0.62 |
| TOTAL | 2,119 | 70.79 | 1.41 |

## Maseru District

| Land Cover Type | Area (km²) | Growing Stock (1,000 tonnes) | Mean Annual Increment (1,000 tonnes) |
|---|---|---|---|
| Escarpment and Riparian Woodland | 153 | 214.81 | 2.29 |
| Escarpment Grassland with Scrub Woodland | 1,678 | 83.90 | 1.68 |
| Highveld and Riparian Grassland | 615 | 0 | 0 |
| Alpine/Sub-Alpine Grassland and Heathland | 1,357 | 0 | 0 |
| Small farm cultivation | 476 | 23.80 | 0.48 |
| TOTAL | 4,279 | 322.51 | 4.45 |

## Mohale's Hoek District

| Land Cover Type | Area (km²) | Growing Stock (1,000 tonnes) | Mean Annual Increment (1,000 tonnes) |
|---|---|---|---|
| Escarpment and Riparian Woodland | 202 | 283.61 | 3.03 |
| Escarpment Grassland with Scrub Woodland | 39 | 1.95 | 0.04 |
| Highveld and Riparian Grassland | 1,831 | 0 | 0 |
| Alpine/Sub-Alpine Grassland and Heathland | 1,105 | 0 | 0 |
| Small farm cultivation | 353 | 17.65 | 0.35 |
| TOTAL | 3,530 | 303.21 | 3.42 |

## Mokhotlong District

| Land Cover Type | Area (km²) | Growing Stock (1,000 tonnes) | Mean Annual Increment (1,000 tonnes) |
|---|---|---|---|
| Escarpment and Riparian Woodland | 208 | 292.03 | 3.12 |
| Escarpment Grassland with Scrub Woodland | 568 | 28.40 | 0.57 |
| Highveld and Riparian Grassland | 106 | 0 | 0 |
| Alpine/Sub-Alpine Grassland and Heathland | 3,046 | 0 | 0 |
| Small farm cultivation | 147 | 7.35 | 0.15 |
| TOTAL | 4,075 | 327.78 | 3.84 |

## Qacha's Nek District

| Land Cover Type | Area (km²) | Growing Stock (1,000 tonnes) | Mean Annual Increment (1,000 tonnes) |
|---|---|---|---|
| Escarpment Grassland with Scrub Woodland | 237 | 11.84 | 0.24 |
| Highveld and Riparian Grassland | 149 | 0 | 0 |
| Alpine/Sub-Alpine Grassland and Heathland | 1,868 | 0 | 0 |
| Small farm cultivation | 95 | 4.75 | 0.09 |
| TOTAL | 2,349 | 16.59 | 0.33 |

## Quthing District

| Land Cover Type | Area (km²) | Growing Stock (1,000 tonnes) | Mean Annual Increment (1,000 tonnes) |
|---|---|---|---|
| Escarpment and Riparian Woodland | 39 | 54.76 | 0.58 |
| Escarpment Grassland with Scrub Woodland | 649 | 32.45 | 0.65 |
| Highveld and Riparian Grassland | 958 | 0 | 0 |
| Alpine/Sub-Alpine Grassland and Heathland | 1,088 | 0 | 0 |
| Small farm cultivation | 182 | 9.10 | 0.18 |
| TOTAL | 2,916 | 96.31 | 1.42 |

## Thaba Tseka District

| Land Cover Type | Area (km²) | Growing Stock (1,000 tonnes) | Mean Annual Increment (1,000 tonnes) |
|---|---|---|---|
| Escarpment and Riparian Woodland | 25 | 35.10 | 0.37 |
| Escarpment Grassland with Scrub Woodland | 854 | 42.70 | 0.85 |
| Highveld and Riparian Grassland | 558 | 0 | 0 |
| Alpine/Sub-Alpine Grassland and Heathland | 2,675 | 0 | 0 |
| Small farm cultivation | 158 | 7.90 | 0.16 |
| TOTAL | 4,270 | 85.70 | 1.38 |

## MALAWI
### Central Region

| Land Cover Type | Area (km²) | Growing Stock (mill t) | Mean Annual Increment (mill t) |
|---|---|---|---|
| Evergreen/Semi-Deciduous Forest and Woodland | 1,045 | 7.44 | 0.24 |
| Seasonal Open-Canopy Miombo Woodland | 6,973 | 13.84 | 0.34 |
| Miombo Woodland with extensive tobacco cultivation | 4,562 | 7.66 | 0.11 |
| Dry, Open-Canopy Miombo Woodland and Cultivation Savannah | 11,110 | 10.49 | 0.27 |
| Mopane Woodland | 650 | 2.40 | 0.07 |
| Swamp Woodland and Grassland | 2,339 | 0 | 0 |
| *Sub-total* | 26,679 | 41.83 | 1.03 |
| RESTRICTED LAND | | | |
| Evergreen/Semi-Deciduous Forest and Woodland | 9,407 | 67.00 | 2.12 |
| Seasonal Open-Canopy Miombo Woodland | 775 | 1.54 | 0.04 |
| Plantations (*Gmelina and Eucalyptus*) | 1,023 | 4.30 | 0.87 |
| Tobacco estates | 6,840 | 11.48 | 0.17 |
| *Sub-total* | 18,045 | 84.30 | 3.20 |
| TOTAL | 44,724 | 126.13 | 4.23 |

## Northern Region

| Land Cover Type | Area (km²) | Growing Stock (mill t) | Mean Annual Increment (mill t) |
|---|---|---|---|
| Evergreen/Semi-Deciduous Forest and Woodland | 10,723 | 76.37 | 2.41 |
| Seasonal Open-Canopy Miombo Woodland | 2,234 | 4.43 | 0.11 |
| Miombo Woodland with extensive tobacco cultivation | 1,642 | 2.76 | 0.04 |
| Dry, Open Canopy Miombo Woodland and Cultivation Savannah | 744 | 0.70 | 0.02 |
| Swamp Woodland and Grassland | 4,874 | 0 | 0 |
| *Sub-total* | 20,217 | 84.26 | 2.58 |
| RESTRICTED LAND | | | |
| Evergreen/Semi-Deciduous Forest and Woodland | 877 | 62.49 | 1.97 |
| Plantations (coniferous) | 3,215 | 13.53 | 2.74 |
| Tobacco estates | 1,642 | 2.76 | 0.04 |
| *Sub-total* | 13,631 | 78.79 | 4.75 |
| TOTAL | 33,848 | 163.05 | 7.33 |

## Southern Region

| Land Cover Type | Area (km²) | Growing Stock (mill t) | Mean Annual Increment (mill t) |
|---|---|---|---|
| Seasonal Open-Canopy Miombo Woodland | 14,715 | 29.21 | 0.72 |
| Miombo Woodland with extensive tobacco cultivation | 969 | 1.63 | 0.02 |
| Dry Open-Canopy Miombo Woodland and Cultivation Savannah | 8,398 | 7.93 | 0.20 |
| Mopane Woodland | 214 | 0.79 | 0.02 |
| Swamp Rice, Tea and Coffee Cultivation | 8,326 | 0 | 0 |
| *Sub-total* | 32,622 | 39.56 | 0.78 |
| RESTRICTED LAND | | | |
| Evergreen/Semi-Deciduous Forest and Woodland | 5,251 | 37.40 | 1.18 |
| Tobacco estates | 969 | 1.63 | 0.02 |
| *Sub-total* | 6,220 | 39.01 | 1.20 |
| TOTAL | 38,842 | 78.57 | 1.98 |

## MOZAMBIQUE
### Cabo Delgado Province

| Land Cover Type | Area (km²) | Growing Stock (mill t) | Mean Annual Increment (mill t) |
|---|---|---|---|
| Evergreen Miombo Woodland and Coastal Forest | 238 | 1.70 | 0.05 |
| Wet Seasonal Forest and Woodland | 25,074 | 178.58 | 5.64 |
| Seasonal Miombo Woodland | 35,820 | 71.10 | 1.76 |
| Dry Miombo Woodland/Wood and Shrub Thicket | 21,492 | 20.29 | 0.52 |
| TOTAL | 82,624 | 271.67 | 7.97 |

### Gaza Province

| Land Cover Type | Area (km²) | Growing Stock (mill t) | Mean Annual Increment (mill t) |
|---|---|---|---|
| Evergreen Miombo Woodland | 6,561 | 46.73 | 1.48 |
| Wet Seasonal Forest and Woodland | 4,080 | 29.06 | 0.92 |
| Seasonal Miombo Woodland | 6,045 | 12.00 | 0.30 |
| Dry Miombo Woodland/Wood and Shrub Thicket | 24,179 | 22.82 | 0.58 |
| Coastal Forest Mosaic | 9,143 | 81.15 | 0.45 |
| Dry Riparian Woodland | 19,267 | 22.06 | 1.06 |
| Degraded Agricultural Land | 6,435 | 15.03 | 0.53 |
| TOTAL | 75,710 | 228.85 | 5.32 |

## Inhambane Province

| Land Cover Type | Area (km²) | Growing Stock (mill t) | Mean Annual Increment (mill t) |
|---|---|---|---|
| Evergreen Miombo Woodland and Coastal Forest | 177 | 1.26 | 0.04 |
| Wet Seasonal Forest and Woodland | 2,660 | 18.94 | 0.60 |
| Seasonal Miombo Woodland | 1,684 | 3.34 | 0.08 |
| Dry Miombo Woodland/Wood and Shrub Thicket | 21,719 | 20.50 | 0.52 |
| Coastal Forest Mosaic | 30,318 | 269.17 | 15.10 |
| Dry Riparian Woodland | 354 | 0.41 | 0.02 |
| Degraded Agricultural Land | 11,702 | 27.34 | 0.97 |
| TOTAL | 68,615 | 340.96 | 17.33 |

## Manica Province

| Land Cover Type | Area (km²) | Growing Stock (mill t) | Mean Annual Increment (mill t) |
|---|---|---|---|
| Wet Seasonal Forest and Woodland | 30,235 | 215.33 | 6.80 |
| Seasonal Miombo Woodland | 17,503 | 34.74 | 0.86 |
| Dry Miombo Woodland/Wood and Shrub Thicket | 10,741 | 10.14 | 0.26 |
| Mopane Woodland | 3,182 | 11.76 | 0.35 |
| TOTAL | 61,661 | 271.97 | 8.27 |

## Maputo Province

| Land Cover Type | Area (km²) | Growing Stock (mill t) | Mean Annual Increment (mill t) |
|---|---|---|---|
| Seasonal Miombo Woodland | 655 | 1.30 | 0.03 |
| Lubombo Hills Woodland | 1,473 | 11.18 | 0.39 |
| Coastal Forest Mosaic | 6,950 | 61.70 | 3.46 |
| Dry Miombo Woodland/Wood and Scrub Thicket | 1,022 | 0.96 | 0.02 |
| Dry Riparian Woodland | 2,917 | 3.34 | 0.16 |
| Degraded Agricultural Land | 655 | 1.53 | 0.05 |
| Lowland Sublittoral Forest and Bushland | 3,638 | 3.43 | 0.09 |
| Littoral Grassland | 9,048 | 0 | 0 |
| TOTAL | 26,358 | 83.45 | 4.21 |

## Nampula Province

| Land Cover Type | Area (km²) | Growing Stock (mill t) | Mean Annual Increment (mill t) |
|---|---|---|---|
| Evergreen Miombo Woodland and Coastal Forest | 32,755 | 233.28 | 7.37 |
| Wet Seasonal Forest and Woodland | 25,196 | 179.45 | 5.67 |
| Seasonal Miombo Woodland | 22,116 | 43.90 | 1.08 |
| Dry Miombo Woodland/Wood Shrub and Thicket | 1,120 | 1.06 | 0.03 |
| Dry Riparian Woodland | 420 | 0.48 | 0.02 |
| TOTAL | 81,607 | 458.17 | 14.17 |

## Niassa Province

| Land Cover Type | Area (km²) | Growing Stock (mill t) | Mean Annual Increment (mill t) |
|---|---|---|---|
| Evergreen Miombo Woodland | 8,413 | 59.92 | 1.89 |
| Wet Seasonal Forest and Woodland | 68,434 | 487.39 | 15.40 |
| Seasonal Miombo Woodland | 38,687 | 76.79 | 1.90 |
| Dry Miombo Woodland/Wood and Shrub Thicket | 13,521 | 12.76 | 0.32 |
| TOTAL | 129,055 | 636.86 | 19.51 |

## Sofala Province

| Land Cover Type | Area (km²) | Growing Stock (mill t) | Mean Annual Increment (mill t) |
|---|---|---|---|
| Evergreen Miombo Woodland and Coastal Forest | 16,258 | 115.79 | 3.66 |
| Wet Seasonal Forest and Woodland | 19,576 | 139.42 | 4.40 |
| Seasonal Miombo Woodland | 18,581 | 36.88 | 0.91 |
| Dry Miombo Woodland/Wood and Shrub Thicket | 11,613 | 10.96 | 0.28 |
| Coastal Forest | 497 | 4.41 | 0.25 |
| Degraded Agricultural land | 497 | 1.16 | 0.04 |
| Littoral Grassland | 995 | 0 | 0 |
| TOTAL | 68,017 | 308.63 | 9.54 |

## Tete Province

| Land Cover Type | Area (km²) | Growing Stock (mill t) | Mean Annual Increment (mill t) |
|---|---|---|---|
| Evergreen Miombo Woodland | 1,266 | 9.02 | 0.28 |
| Wet Seasonal Forest and Woodland | 13,303 | 94.74 | 2.99 |
| Seasonal Miombo Woodland | 38,010 | 75.45 | 1.86 |
| Dry Miombo Woodland/Wood and Shrub Thicket | 39,784 | 37.56 | 0.95 |
| Mopane Woodland | 8,362 | 30.91 | 0.91 |
| TOTAL | 100,725 | 247.68 | 6.99 |

## Zambezia Province

| Land Cover Type | Area (km²) | Growing Stock (mill t) | Mean Annual Increment (mill t) |
|---|---|---|---|
| Evergreen Miombo Woodland and Coastal Forest | 55,334 | 394.09 | 12.45 |
| Wet Seasonal Forest and Woodland | 34,583 | 246.30 | 7.78 |
| Seasonal Miombo Woodland | 11,738 | 23.30 | 0.58 |
| Dry Miombo Woodland/Wood and Shrub Thicket | 3,354 | 3.17 | 0.08 |
| TOTAL | 105,009 | 666.86 | 20.89 |

## SWAZILAND
### Hhohho District

| Land Cover Type | Area (km²) | Growing Stock (mill t) | Mean Annual Increment (mill t) |
|---|---|---|---|
| Highveld Forest | 851 | 1.19 | 0.01 |
| Middleveld Dense Woodland and Bushland | 521 | 1.22 | 0.04 |
| Dense Plantation Stands | 625 | 2.47 | 0.32 |
| Middleveld Sparse Bushland and Woodland | 956 | 1.37 | 0.06 |
| Middleveld Open Grassy Savannah | 353 | 0.25 | 0.01 |
| FOREST/GAME RESERVES | | | |
| Sparse Bushland and Woodland | 238 | 0.34 | 0.01 |
| TOTAL | 3,544 | 6.84 | 0.45 |

### Lubombo District

| Land Cover Type | Area (km²) | Growing Stock (mill t) | Mean Annual Increment (mill t) |
|---|---|---|---|
| Irrigated Agriculture | 1,465 | 0 | 0 |
| Lowveld Dense Woodland and Bushland | 1,168 | 2.73 | 0.10 |
| Lubombo Hills Woodland | 64 | 0.49 | 0.02 |
| Lowveld Sparse Bushland and Woodland | 851 | 1.22 | 0.05 |
| Lowveld Open Grassy Savannah | 2,160 | 1.54 | 0.06 |
| FOREST/GAME RESERVES | | | |
| Lowland Dense Woodland and Bushland | 293 | 0.68 | 0.02 |
| TOTAL | 6,001 | 6.66 | 0.25 |

## Manzini District

| Land Cover Type | Area (km²) | Growing Stock (mill t) | Mean Annual Increment (mill t) |
|---|---|---|---|
| Highveld Forest | 96 | 0.13 | 0.01 |
| Middleveld Dense Woodland and Bushland | 421 | 0.98 | 0.03 |
| Dense Plantation Stands | 424 | 1.67 | 0.22 |
| Middleveld Sparse Bushland and Woodland | 189 | 0.27 | 0.01 |
| Middleveld Open Grassy Savannah | 2,022 | 1.44 | 0.06 |
| FOREST/GAME RESERVES | | | |
| Highveld Forest | 865 | 1.21 | 0.01 |
| TOTAL | 4,017 | 5.70 | 0.33 |

## Shiselweni District

| Land cover type | Area (km²) | Growing Stock (mill t) | Mean Annual Increment (mill t) |
|---|---|---|---|
| Highveld Forest | 654 | 0.92 | 0.01 |
| Lowveld/Middleveld Dense Woodland and Bushland | 905 | 2.11 | 0.08 |
| Wattle and *Eucalyptus* Plantations | 114 | 0.48 | 0.10 |
| Lowveld/Middleveld Sparse Bushland and Woodland | 79 | 0.11 | 0.01 |
| Open Grassy Savannah | 2,001 | 1.42 | 0.06 |
| FOREST/GAME RESERVES | | | |
| Wattle and *Eucalyptus* Plantations | 49 | 0.21 | 0.04 |
| TOTAL | 3,802 | 5.25 | 0.30 |

## TANZANIA
## Arusha District

| Land Cover Type | Area (km²) | Growing Stock (mill t) | Mean Annual Increment (mill t) |
|---|---|---|---|
| Semi-Arid Dry Steppe | 3,069 | 3.51 | 0.17 |
| Semi-Arid Steppe | 21,884 | 20.66 | 0.53 |
| Wet Seasonal Miombo Woodland | 8,184 | 58.29 | 1.84 |
| FOREST/GAME RESERVES | | | |
| Semi-Arid Dry Steppe | 27,620 | 31.62 | 1.52 |
| Semi-Arid Steppe | 17,898 | 16.90 | 0.43 |
| Wet Seasonal Miombo Woodland | 2,046 | 14.57 | 0.46 |
| TOTAL | 80,701 | 145.54 | 4.95 |

## Kilimanjaro District

| Land Cover Type | Area (km²) | Growing Stock (mill t) | Mean Annual Increment (mill t) |
|---|---|---|---|
| Semi-Arid Dry Steppe | 6,696 | 7.67 | 0.37 |
| Semi-Arid Steppe | 33 | 0.03 | <0.01 |
| Wet Seasonal Miombo Woodland | 746 | 5.31 | 0.17 |
| FOREST/GAME RESERVES | | | |
| Semi-Arid Dry Steppe | 4,464 | 5.11 | 0.25 |
| Semi-Arid Steppe | 78 | 0.07 | <0.01 |
| Wet Seasonal Miombo Woodland | 83 | 0.59 | 0.02 |
| TOTAL | 12,100 | 18.78 | 0.81 |

## Dodoma District

| Land Cover Type | Area (km²) | Growing Stock (mill t) | Mean Annual Increment (mill t) |
|---|---|---|---|
| Semi-Arid Dry Steppe | 2,792 | 3.20 | 0.15 |
| Semi-Arid Steppe | 24,188 | 22.83 | 0.58 |
| Dry Miombo Woodland | 1,875 | 3.72 | 0.09 |
| Coastal Forest Mosaic | 1,875 | 16.65 | 0.93 |
| Wet Seasonal Miombo Woodland | 2,932 | 20.88 | 0.6 |
| Cleared Miombo Woodland | 1,396 | 4.68 | 0.14 |
| FOREST/GAME RESERVES | | | |
| Semi-Arid Steppe | 2,688 | 2.54 | 0.06 |
| Wet-Seasonal Miombo Woodland | 1,256 | 8.95 | 0.28 |
| TOTAL | 39,002 | 83.45 | 2.89 |

## Mara District

| Land Cover Type | Area (km²) | Growing Stock (mill t) | Mean Annual Increment (mill t) |
|---|---|---|---|
| Semi-Arid Dry Steppe | 4,930 | 5.64 | 0.27 |
| Semi-Arid Steppe | 234 | 0.22 | 0.01 |
| Wet Seasonal Miombo Woodland | 2,094 | 14.91 | 0.47 |
| FOREST/GAME RESERVES | | | |
| Semi-Arid Dry Steppe | 11,504 | 13.17 | 0.63 |
| Semi-Arid Steppe | 938 | 0.89 | 0.02 |
| TOTAL | 19,700 | 34.83 | 1.40 |

## Iringa District

| Land Cover Type | Area (km²) | Growing Stock (mill t) | Mean Annual Increment (mill t) |
|---|---|---|---|
| Semi-Arid Dry Steppe | 5,058 | 5.79 | 0.28 |
| Semi-Arid Steppe | 7,435 | 7.02 | 0.18 |
| Dry Miombo Woodland | 292 | 0.58 | 0.01 |
| Wet Miombo Woodland | 875 | 6.23 | 0.20 |
| Wet Seasonal Miombo Woodland | 24,686 | 175.82 | 5.55 |
| Cleared Miombo Woodland | 2,023 | 6.79 | 0.20 |
| FOREST/GAME RESERVES | | | |
| Semi-Arid Steppe | 9,435 | 8.91 | 0.23 |
| Wet Seasonal Miombo Woodland | 2,743 | 19.54 | 0.62 |
| TOTAL | 52,549 | 230.68 | 7.27 |

## Kagera District

| Land Cover Type | Area (km²) | Growing Stock (mill t) | Mean Annual Increment (mill t) |
|---|---|---|---|
| Semi-Arid Dry Steppe | 3,693 | 4.23 | 0.20 |
| Semi-Arid Steppe | 722 | 0.73 | 0.02 |
| Dry Miombo Woodland | 8,620 | 17.11 | 0.42 |
| Wet Miombo Woodland | 4,524 | 32.22 | 1.02 |
| Wet Seasonal Miombo Woodland | 3,968 | 28.26 | 0.89 |
| FOREST/GAME RESERVES | | | |
| Dry Miombo Woodland | 1,521 | 3.02 | 0.07 |
| Wet Miombo Woodland | 503 | 3.58 | 0.11 |
| TOTAL | 23,601 | 89.15 | 2.73 |

## Mtwara District

| Land Cover Type | Area (km²) | Growing Stock (mill t) | Mean Annual Increment (mill t) |
|---|---|---|---|
| Semi-Arid Dry Steppe | 661 | 0.76 | 0.04 |
| Semi-Arid Steppe | 3,967 | 3.74 | 0.10 |
| Coastal Forest Mosaic | 1,543 | 13.70 | 0.77 |
| Wet Seasonal Miombo Woodland | 14,281 | 101.71 | 3.21 |
| FOREST/GAME RESERVES | | | |
| Wet Seasonal Miombo Woodland | 1,587 | 11.30 | 0.36 |
| TOTAL | 22,039 | 131.21 | 4.48 |

## Kigoma District

| Land Cover Type | Area (km²) | Growing Stock (mill t) | Mean Annual Increment (mill t) |
|---|---|---|---|
| Semi-Arid Steppe | 3,743 | 3.53 | 0.09 |
| Dry Miombo Woodland | 8,167 | 16.21 | 0.40 |
| Wet Miombo Woodland | 7,980 | 56.83 | 0.80 |
| Wet Seasonal Miombo Woodland | 1,021 | 7.27 | 0.23 |
| Cleared Miombo Woodland | 6,296 | 21.12 | 0.64 |
| FOREST/GAME RESERVES | | | |
| Dry Miombo Woodland | 2,042 | 4.05 | 0.10 |
| Wet Miombo Woodland | 3,420 | 24.36 | 0.77 |
| Wet Seasonal Miombo Woodland | 1,531 | 10.90 | 0.34 |
| TOTAL | 34,200 | 144.29 | 3.37 |

## Lindi District

| Land Cover Type | Area (km²) | Growing Stock (mill t) | Mean Annual Increment (mill t) |
|---|---|---|---|
| Semi-Arid Dry Steppe | 1,653 | 1.89 | 0.09 |
| Semi-Arid Steppe | 1,653 | 1.56 | 0.04 |
| Wet Miombo Woodland | 166 | 1.18 | 0.04 |
| Dry Miombo Woodland | 154 | 0.31 | 0.01 |
| Wet Seasonal Miombo Woodland | 25,345 | 180.51 | 5.70 |
| Cleared Miombo Woodland | 815 | 2.73 | 0.08 |
| FOREST/GAME RESERVES | | | |
| Wet Miombo Woodland | 1,487 | 10.59 | 0.33 |
| Coastal Forest Mosaic | 297 | 2.64 | 0.15 |
| Wet Seasonal Miombo Woodland | 25,345 | 180.51 | 5.70 |
| TOTAL | 57,610 | 388.09 | 12.49 |

## Mbeya District

| Land Cover Type | Area (km²) | Growing Stock (mill t) | Mean Annual Increment (mill t) |
|---|---|---|---|
| Semi-Arid Dry Steppe | 1,929 | 2.21 | 0.11 |
| Semi-Arid Steppe | 24,311 | 22.95 | 0.58 |
| Dry Miombo Woodland | 4,438 | 8.81 | 0.22 |
| Wet Miombo Woodland | 772 | 5.50 | 0.17 |
| Wet Seasonal Miombo Woodland | 13,506 | 96.19 | 3.04 |
| Cleared Miombo Woodland | 30,871 | 103.57 | 3.12 |
| FOREST/GAME RESERVES | | | |
| Dry Miombo Woodland | 4,438 | 8.81 | 0.22 |
| Wet Miombo Woodland | 386 | 2.75 | 0.09 |
| TOTAL | 80,651 | 250.79 | 7.55 |

## Morogoro District

| Land Cover Type | Area (km²) | Growing Stock (mill t) | Mean Annual Increment (mill t) |
|---|---|---|---|
| Semi-Arid Dry Steppe | 1,503 | 1.72 | 0.08 |
| Semi-Arid Steppe | 2,781 | 2.63 | 0.07 |
| Dry Miombo Woodland | 1,009 | 2.00 | 0.05 |
| Wet Miombo Woodland | 3,364 | 23.96 | 0.76 |
| Coastal Forest Mosaic | 841 | 7.47 | 0.42 |
| Wet Seasonal Miombo Woodland | 37,676 | 268.33 | 8.48 |
| Cleared Miombo Woodland | 2,534 | 8.50 | 0.26 |
| FOREST/GAME RESERVES | | | |
| Wet Miombo Woodland | 5,046 | 35.94 | 1.14 |
| Wet Seasonal Miombo Woodland | 16,147 | 115.00 | 3.63 |
| TOTAL | 70,901 | 465.54 | 14.89 |

## Pwani District

| Land Cover Type | Area (km²) | Growing Stock (mill t) | Mean Annual Increment (mill t) |
|---|---|---|---|
| Semi-Arid Dry Steppe | 5,916 | 6.77 | 0.33 |
| Wet Miombo Woodland | 1,183 | 8.43 | 0.27 |
| Coastal Forest Mosaic | 9,269 | 82.29 | 4.62 |
| Wet Seasonal Miombo Woodland | 13,667 | 97.34 | 3.08 |
| Cleared Miombo Woodland | 296 | 0.99 | 0.03 |
| FOREST/GAME RESERVES | | | |
| Wet Seasonal Miombo Woodland | 1,519 | 10.82 | 0.34 |
| TOTAL | 31,850 | 206.64 | 8.65 |

## Mwanza District

| Land Cover Type | Area (km²) | Growing Stock (mill t) | Mean Annual Increment (mill t) |
|---|---|---|---|
| Semi-Arid Dry Steppe | 6,341 | 7.26 | 0.35 |
| Semi-Arid Steppe | 1,963 | 1.85 | 0.05 |
| Dry Miombo Woodland | 3,261 | 6.47 | 0.16 |
| Wet Miombo Woodland | 3,019 | 21.50 | 0.68 |
| Wet Seasonal Miombo Woodland | 604 | 4.30 | 0.14 |
| FOREST/GAME RESERVES | | | |
| Dry Miombo Woodland | 362 | 0.72 | 0.02 |
| TOTAL | 15,550 | 42.10 | 1.40 |

## Rukwa District

| Land Cover Type | Area (km²) | Growing Stock (mill t) | Mean Annual Increment (mill t) |
|---|---|---|---|
| Semi-Arid Steppe | 5,070 | 4.79 | 0.12 |
| Dry Miombo Woodland | 1,613 | 3.20 | 0.08 |
| Wet Seasonal Miombo Woodland | 8,297 | 59.09 | 1.87 |
| Cleared Miombo Woodland | 21,895 | 73.46 | 2.21 |
| FOREST/GAME RESERVES | | | |
| Dry Miombo Woodland | 14,520 | 28.82 | 0.71 |
| Wet Miombo Woodland | 14,520 | 103.41 | 3.27 |
| Wet Seasonal Miombo Woodland | 2,074 | 14.77 | 0.47 |
| TOTAL | 67,989 | 287.54 | 8.73 |

**Ruvuma District**

| Land Cover Type | Area (km²) | Growing Stock (mill t) | Mean Annual Increment (mill t) |
|---|---|---|---|
| Semi-Arid Dry Steppe | 417 | 0.48 | 0.02 |
| Semi-Arid Steppe | 1,398 | 1.32 | 0.03 |
| Dry Miombo Woodland | 2,107 | 4.18 | 0.10 |
| Wet Miombo Woodland | 12,957 | 92.28 | 2.92 |
| Wet Seasonal Miombo Woodland | 37,557 | 267.48 | 8.45 |
| FOREST/GAME RESERVES | | | |
| Wet Miombo Woodland | 1,437 | 10.26 | 0.32 |
| Wet Seasonal Miombo Woodland | 4,173 | 29.72 | 0.94 |
| TOTAL | 60,049 | 405.72 | 12.79 |

## Shinyanga District

| Land Cover Type | Area (km²) | Growing Stock (mill t) | Mean Annual Increment (mill t) |
|---|---|---|---|
| Semi-Arid Dry Steppe | 6,924 | 7.93 | 0.38 |
| Semi-Arid Steppe | 14,809 | 13.98 | 0.36 |
| Dry Miombo Woodland | 12,809 | 25.43 | 0.63 |
| Wet Miombo Woodland | 1,077 | 7.67 | 0.24 |
| Wet Seasonal Miombo Woodland | 115 | 0.82 | 0.03 |
| FOREST/GAME RESERVES | | | |
| Semi-Arid Dry Steppe | 1,731 | 1.98 | 0.01 |
| Semi-Arid Steppe | 6,347 | 5.99 | 0.15 |
| Dry Miombo Woodland | 1,423 | 2.83 | 0.07 |
| Wet Miombo Woodland | 1,077 | 7.67 | 0.24 |
| Wet Seasonal Miombo Woodland | 1,038 | 7.39 | 0.23 |
| TOTAL | 47,350 | 81.68 | 2.43 |

## Singida District

| Land Cover Type | Area (km²) | Growing Stock (mill t) | Mean Annual Increment (mill t) |
|---|---|---|---|
| Semi-Arid Dry Steppe | 2,443 | 2.80 | 0.13 |
| Semi-Arid Steppe | 23,889 | 22.55 | 0.57 |
| Dry Miombo Woodland | 5,288 | 10.50 | 0.26 |
| Wet Seasonal Miombo Woodland | 1,412 | 10.06 | 0.32 |
| Cleared Miombo Woodland | 814 | 2.73 | 0.08 |
| FOREST/GAME RESERVES | | | |
| Semi-Arid Steppe | 5,972 | 5.64 | 0.14 |
| Dry Miombo Woodland | 7,932 | 15.75 | 0.39 |
| TOTAL | 47,750 | 70.03 | 1.89 |

## Tabora District

| Land Cover Type | Area (km²) | Growing Stock (mill t) | Mean Annual Increment (mill t) |
|---|---|---|---|
| Semi-Arid Steppe | 2,420 | 2.28 | 0.06 |
| Dry Miombo Woodland | 13,921 | 27.63 | 0.68 |
| Wet Miombo Woodland | 3,861 | 27.50 | 0.87 |
| Wet Seasonal Miombo Woodland | 853 | 6.08 | 0.19 |
| Cleared Miombo Woodland | 3,088 | 10.36 | 0.31 |
| FOREST/GAME RESERVES | | | |
| Semi-Arid Dry Steppe | 1,890 | 2.16 | 0.10 |
| Semi-Arid Steppe | 1,037 | 0.98 | 0.02 |
| Dry Miombo Woodland | 20,881 | 41.45 | 1.02 |
| Wet Miombo Woodland | 3,860 | 27.49 | 0.87 |
| TOTAL | 51,811 | 145.93 | 4.12 |

## Tanga District

| Land Cover Type | Area (km²) | Growing Stock (mill t) | Mean Annual Increment (mill t) |
|---|---|---|---|
| Semi-Arid Dry Steppe | 582 | 0.67 | 0.03 |
| Semi-Arid Steppe | 427 | 0.40 | 0.01 |
| Wet Miombo Woodland | 427 | 3.04 | 0.10 |
| Coastal Forest Mosaic | 582 | 5.17 | 0.29 |
| Wet Seasonal Miombo Woodland | 13,959 | 99.42 | 3.14 |
| FOREST/GAME RESERVES | | | |
| Semi-Arid Dry Steppe | 5,235 | 5.99 | 0.29 |
| Wet Seasonal Miombo Woodland | 3,490 | 24.86 | 0.79 |
| TOTAL | 24,701 | 139.54 | 4.65 |

**ZAMBIA**
**Central Province**

| Land Cover Type | Area (km²) | Growing Stock (mill t) | Mean Annual Increment (mill t) |
|---|---|---|---|
| Wet Miombo Woodland | 30,023 | 213.83 | 6.76 |
| Kalahari Woodland | 1,140 | 1.91 | 0.06 |
| Dry Miombo and Munga Woodland | 6,610 | 6.24 | 0.16 |
| Seasonal Miombo Woodland | 17,970 | 35.67 | 0.88 |
| Dry Evergreen Woodland | 877 | 6.25 | 0.20 |
| Degraded Miombo Woodland | 7,188 | 24.12 | 0.73 |
| Mopane Woodland | 4,405 | 16.29 | 0.48 |
| Scrub Woodland | 219 | 0.51 | 0.02 |
| Swamp and Lake Vegetation | 3,287 | 0 | 0 |
| FOREST/GAME RESERVES | | | |
| Dry Miombo and Munga Woodland | 1,652 | 1.56 | 0.04 |
| Degraded Miombo Woodland | 7,188 | 24.12 | 0.73 |
| TOTAL | 80,560 | 330.49 | 10.06 |

**Copperbelt Province**

| Land Cover Type | Area (km²) | Growing Stock (mill t) | Mean Annual Increment (mill t) |
|---|---|---|---|
| Wet Miombo Woodland | 12,880 | 91.73 | 2.90 |
| Kalahari Woodland | 17 | 0.03 | 0 |
| Seasonal Miombo Woodland | 8,951 | 17.77 | 0.44 |
| Degraded Miombo Woodland | 8,951 | 30.03 | 0.90 |
| Swamp and Lake Vegetation | 166 | 0 | 0 |
| TOTAL | 30,965 | 139.56 | 4.24 |

**Luapula Province**

| Land Cover Type | Area (km²) | Growing Stock (mill t) | Mean Annual Increment (mill t) |
|---|---|---|---|
| Wet Miombo Woodland | 20,387 | 145.20 | 4.59 |
| Kalahari Woodland | 106 | 0.18 | 0.01 |
| Seasonal Miombo Woodland | 10,570 | 20.98 | 0.52 |
| Dry Evergreen Woodland | 106 | 0.75 | 0.02 |
| Degraded Miombo Woodland | 2,973 | 9.97 | 0.30 |
| Swamp and Lake Vegetation | 6,644 | 0 | 0 |
| FOREST/GAME RESERVE | | | |
| Wet Miombo Woodland | 2,263 | 16.12 | 0.51 |
| Degraded Miombo Woodland | 6,933 | 23.26 | 0.70 |
| TOTAL | 49,982 | 216.46 | 6.65 |

## Eastern Province

| Land Cover Type | Area (km²) | Growing Stock (mill t) | Mean Annual Increment (mill t) |
|---|---|---|---|
| Wet Miombo Woodland | 11,482 | 81.77 | 2.58 |
| Kalahari Woodland | 2,625 | 4.41 | 0.13 |
| Dry Miombo and Munga Woodland | 5,250 | 4.96 | 0.13 |
| Seasonal Miombo Woodland | 27,584 | 54.75 | 1.35 |
| Dry Evergreen Woodland | 175 | 1.25 | 0.04 |
| Degraded Miombo Woodland | 5,828 | 19.55 | 0.59 |
| Mopane Woodland | 5,250 | 19.41 | 0.57 |
| Swamp and Lake Vegetation | 350 | 0 | 0 |
| FOREST/GAME RESERVES | | | |
| Wet Miombo Woodland | 2,869 | 20.43 | 0.65 |
| Seasonal Miombo Woodland | 6,892 | 13.68 | 0.34 |
| TOTAL | 68,305 | 220.21 | 6.38 |

## Lusaka Province

| Land Cover Type | Area (km²) | Growing Stock (mill t) | Mean Annual Increment (mill t) |
|---|---|---|---|
| Wet Miombo Woodland | 9,862 | 70.24 | 2.22 |
| Kalahari Woodland | 986 | 1.66 | 0.05 |
| Dry Miombo and Munga Woodland | 7,889 | 7.45 | 0.19 |
| Seasonal Miombo Woodland | 2,126 | 4.22 | 0.10 |
| Degraded Miombo Woodland | 2,126 | 7.13 | 0.21 |
| Mopane Woodland | 7,013 | 25.93 | 0.76 |
| Swamp and Lake Vegetation | 4,383 | 0 | 0 |
| TOTAL | 34,385 | 116.63 | 3.53 |

## Northern Province

| Land Cover Type | Area (km²) | Growing Stock (mill t) | Mean Annual Increment (mill t) |
|---|---|---|---|
| Wet Miombo Woodland | 48,203 | 343.30 | 10.85 |
| Kalahari Woodland | 4,789 | 8.04 | 0.23 |
| Dry Miombo and Munga Woodland | 1,916 | 1.81 | 0.05 |
| Seasonal Miombo Woodland | 8,302 | 16.48 | 0.41 |
| Dry Evergreen Woodland | 1,277 | 9.09 | 0.29 |
| Degraded Miombo Woodland | 52,156 | 174.98 | 5.27 |
| Mopane Woodland | 273 | 1.01 | 0.03 |
| Scrub Woodland | 213 | 0.50 | 0.02 |
| FOREST/GAME RESERVES | | | |
| Wet Miombo Woodland | 12,043 | 85.77 | 2.71 |
| TOTAL | 129,172 | 640.98 | 19.86 |

## North-Western Province

| Land Cover Type | Area (km²) | Growing Stock (mill t) | Mean Annual Increment (mill t) |
|---|---|---|---|
| Wet Miombo Woodland | 46,544 | 331.49 | 10.47 |
| Kalahari Woodland | 7,925 | 13.31 | 0.39 |
| Seasonal Miombo Woodland | 30,689 | 60.92 | 1.50 |
| Dry Evergreen Woodland | 1,132 | 8.06 | 0.25 |
| Degraded Miombo Woodland | 13,207 | 44.31 | 1.33 |
| Mopane Woodland | 503 | 1.86 | 0.05 |
| Scrub Woodland | 754 | 1.76 | 0.06 |
| Swamp and Lake Vegetation | 9,056 | 0 | 0 |
| FOREST/GAME RESERVE | | | |
| Wet Miombo Woodland | 11,063 | 78.79 | 2.49 |
| Kalahari Woodland | 3,395 | 5.70 | 0.17 |
| TOTAL | 124,269 | 546.20 | 16.73 |

## Southern Province

| Land Cover Type | Area (km²) | Growing Stock (mill t) | Mean Annual Increment (mill t) |
|---|---|---|---|
| Wet Miombo Woodland | 2,310 | 16.45 | 0.52 |
| Kalahari Woodland | 7,701 | 12.93 | 0.38 |
| Dry Miombo and Munga Woodland | 25,608 | 24.17 | 0.61 |
| Seasonal Miombo Woodland | 5,698 | 11.31 | 0.28 |
| Dry Evergreen Woodland | 3,080 | 21.94 | 0.69 |
| Degraded Miombo Woodland | 1,720 | 5.77 | 0.17 |
| Mopane Woodland | 25,412 | 93.95 | 2.77 |
| Swamp and Lake Vegetation | 8,598 | 0 | 0 |
| FOREST/GAME RESERVES | | | |
| Dry Miombo and Munga Woodland | 1,345 | 1.27 | 0.03 |
| Mopane Woodland | 2,824 | 10.44 | 0.31 |
| TOTAL | 84,296 | 198.23 | 5.76 |

**Western Province**

| Land Cover Type | Area (km²) | Growing Stock (mill t) | Mean Annual Increment (mill t) |
|---|---|---|---|
| Wet Miombo Woodland | 14,013 | 99.80 | 3.15 |
| Kalahari Woodland | 50,528 | 84.84 | 2.48 |
| Dry Miombo and Munga Woodland | 2,815 | 2.66 | 0.07 |
| Seasonal Miombo Woodland | 6,933 | 13.76 | 0.34 |
| Dry Evergreen Woodland | 3,151 | 22.44 | 0.71 |
| Degraded Miombo Woodland | 1,891 | 6.34 | 0.19 |
| Mopane Woodland | 23,320 | 86.21 | 2.54 |
| Scrub Woodland | 8,614 | 20.12 | 0.71 |
| Swamp and Lake Vegetation | 13,656 | 0 | 0 |
| TOTAL | 124,921 | 336.18 | 10.19 |

## ZIMBABWE
**Manicaland Province**

| Land Cover Type | Area (km²) | Growing Stock (mill t) | Mean Annual Increment (mill t) |
|---|---|---|---|
| Dense Savannah Woodland | 1,387 | 9.88 | 0.31 |
| Open Savannah and *Baikiaea* Woodland/Montane Vegetation | 16,842 | 119.95 | 3.79 |
| Dry Savannah Woodland | 2,052 | 1.94 | 0.05 |
| Mopane Woodland and Escarpment Thicket | 849 | 3.14 | 0.09 |
| Dry Bushy Savannah | 5,307 | 12.40 | 0.44 |
| Wooded Grassland | 2,674 | 3.06 | 0.15 |
| Intensive Commercial Agricultural Land | 2,788 | 0 | 0 |
| FOREST/GAME RESERVES | | | |
| Open Savannah and *Baikiaea* Woodland/Montane Vegetation | 2,972 | 21.17 | 0.67 |
| TOTAL | 34,871 | 171.54 | 5.50 |

## Mashonaland Central Province

| Land Cover Type | Area (km²) | Growing Stock (mill t) | Mean Annual Increment (mill t) |
|---|---|---|---|
| Dense Savannah Woodland | 603 | 4.29 | 0.14 |
| Open Savannah and *Baikiaea* Woodland/Montane Vegetation | 1,368 | 9.74 | 0.31 |
| Dry Savannah Woodland | 9,638 | 9.10 | 0.23 |
| Mopane Woodland and Escarpment Thicket | 6,024 | 22.27 | 0.66 |
| Wooded Grassland | 411 | 0.47 | 0.02 |
| Intensive Commercial Agricultural Land | 9,241 | 0 | 0 |
| TOTAL | 27,285 | 45.87 | 1.36 |

## Mashonaland East Province

| Land Cover Type | Area (km²) | Growing Stock (mill t) | Mean Annual Increment (mill t) |
|---|---|---|---|
| Dense Savannah Woodland | 1,073 | 7.64 | 0.24 |
| Open Savannah and *Baikiaea* Woodland/Montane Vegetation | 9,669 | 68.86 | 2.18 |
| Seasonal Savannah Woodland | 243 | 0.48 | 0.01 |
| Dry Savannah Woodland | 4,844 | 4.57 | 0.12 |
| Mopane Woodland and Escarpment Thicket | 586 | 2.17 | 0.06 |
| Dry Bushy Savannah | 2,035 | 4.75 | 0.17 |
| Wooded Grassland | 846 | 0.97 | 0.05 |
| Intensive Commercial Agriculture Land | 3,352 | 0 | 0 |
| FOREST/GAME RESERVES | | | |
| Open Savannah and *Baikiaea* Woodland/Montane Vegetation | 1,074 | 7.65 | 0.24 |
| Dry Savannah Woodland | 1,211 | 1.14 | 0.03 |
| TOTAL | 24,933 | 98.23 | 3.10 |

## Mashonaland West Province

| Land Cover Type | Area (km²) | Growing Stock (mill t) | Mean Annual Increment (mill t) |
|---|---|---|---|
| Dense Savannah Woodland | 383 | 2.73 | 0.09 |
| Open Savannah and *Baikiaea* Woodland/Montane Vegetation | 6,237 | 44.42 | 1.40 |
| Seasonal Savannah Woodland | 1,646 | 3.27 | 0.08 |
| Dry Savannah Woodland | 12,434 | 11.74 | 0.30 |
| Mopane Woodland and Escarpment Thicket | 20,605 | 76.18 | 2.25 |
| Wooded Grassland | 266 | 0.30 | 0.01 |
| Intensive Commercial Agricultural Land | 3,483 | 0 | 0 |
| FOREST/GAME RESERVES | | | |
| Dense Savannah Woodland | 383 | 2.73 | 0.09 |
| Open Savannah and *Baikiaea* Woodland/Montane Vegetation | 693 | 4.94 | 0.16 |
| Seasonal Savannah Woodland | 183 | 0.36 | 0.01 |
| Dry Savannah Woodland | 5,329 | 5.03 | 0.13 |
| Mopane Woodland and Escarpment Thicket | 8,826 | 32.63 | 0.96 |
| TOTAL | 60,468 | 114.33 | 5.48 |

## Masvingo Province

| Land Cover Type | Area (km²) | Growing Stock (mill t) | Mean Annual Increment (mill t) |
|---|---|---|---|
| Dense Savannah Woodland | 4,537 | 32.31 | 1.02 |
| Open Savannah and *Baikiaea* Woodland/Montane Vegetation | 579 | 4.12 | 0.13 |
| Seasonal Savannah Woodland | 214 | 0.42 | 0.01 |
| Dry Savannah Woodland | 2,338 | 2.21 | 0.06 |
| Mopane Woodland and Escarpment Thicket | 1,608 | 5.94 | 0.18 |
| Dry Bushy Savannah | 24,360 | 56.90 | 2.02 |
| Intensive Commercial Agricultural Land | 5,876 | 0 | 0 |
| FOREST/GAME RESERVES | | | |
| Dense Savannah Woodland | 1,141 | 8.13 | 0.26 |
| Open Savannah and *Baikiaea* Woodland/Montane Vegetation | 64 | 0.46 | 0.01 |
| Seasonal Savannah Woodland | 858 | 1.70 | 0.04 |
| Dry Bushy Savannah | 2,707 | 6.32 | 0.22 |
| TOTAL | 44,282 | 118.52 | 3.95 |

**Matabeleland North Province**

| Land Cover Type | Area (km²) | Growing Stock (mill t) | Mean Annual Increment (mill t) |
|---|---|---|---|
| Dense Savannah Woodland | 2,462 | 17.53 | 0.55 |
| Open Savannah and *Baikiaea* Woodland/Montane Vegetation | 18,405 | 131.08 | 4.14 |
| Dry Savannah Woodland | 5,062 | 4.78 | 0.12 |
| Mopane Woodland and Escarpment Thicket | 11,328 | 41.88 | 1.23 |
| Dry Bushy Savannah | 3,941 | 9.21 | 0.33 |
| Wooded Grassland | 2,080 | 2.38 | 0.11 |
| Intensive Commercial Agricultural Land | 1,409 | 0 | 0 |
| FOREST/GAME RESERVES | | | |
| Dense Savannah Woodland | 2,462 | 17.53 | 0.55 |
| Open Savannah and *Baikiaea* Woodland/Montane Vegetation | 18,398 | 131.03 | 4.14 |
| Mopane Woodland and Escarpment Thicket | 7,552 | 27.92 | 0.82 |
| Dry Bushy Savannah | 437 | 1.02 | 0.04 |
| TOTAL | 73,536 | 384.36 | 12.03 |

## Matabeleland South Province

| Land Cover Type | Area (km²) | Growing Stock (mill t) | Mean Annual Increment (mill t) |
|---|---|---|---|
| Open Savannah and *Baikiaea* Woodland/Montane Vegetation | 6,155 | 43.84 | 1.38 |
| Mopane Woodland and Escarpment Thicket | 3,016 | 11.15 | 0.33 |
| Dry Bushy Savannah | 50,401 | 117.74 | 4.18 |
| Degraded Savannah | 2,380 | 1.69 | 0.07 |
| Wooded Grassland | 138 | 0.16 | 0.01 |
| Intensive Commercial Agricultural Land | 3,214 | 0 | 0 |
| FOREST/GAME RESERVES | | | |
| Open Savannah and *Baikiaea* Woodland/Montane Vegetation | 1,086 | 7.73 | 0.24 |
| TOTAL | 66,390 | 182.31 | 6.21 |

## Midlands Province

| Land Cover Type | Area (km²) | Growing Stock (mill t) | Mean Annual Increment (mill t) |
|---|---|---|---|
| Dense Savannah Woodland | 896 | 6.38 | 0.20 |
| Open Savannah and *Baikiaea* Woodland/Mopane Vegetation | 32,536 | 231.72 | 7.32 |
| Dry Savannah Woodland | 4,810 | 4.54 | 0.12 |
| Mopane Woodland and Escarpment Thicket | 7,764 | 28.70 | 0.85 |
| Dry Bushy Savannah | 2,468 | 5.77 | 0.20 |
| Wooded Grassland | 238 | 0.27 | 0.01 |
| Intensive Commercial Agricultural Land | 4,555 | 0 | 0 |
| FOREST/GAME RESERVES | | | |
| Dense Savannah Woodland | 3,580 | 25.50 | 0.81 |
| Open Savannah and *Baikaiea* Woodland/Montane Vegetation | 1,712 | 12.19 | 0.39 |
| Mopane Woodland and Escarpment Thicket | 407 | 1.50 | 0.04 |
| TOTAL | 58,966 | 316.57 | 9.94 |

# Index